2019 SQA Specimen and Past Papers with Answers

National 5
CHEMISTRY

2018 & 2019 Exams
and 2017 Specimen Question Paper

HODDER
GIBSON
AN HACHETTE UK COMPANY

This book contains the official SQA 2018 and 2019 Exams, and the 2017 Specimen Question Paper for National 5 Chemistry, with associated SQA-approved answers modified from the official marking instructions that accompany the paper.

In addition the book contains study skills advice. This has been specially commissioned by Hodder Gibson, and has been written by experienced senior teachers and examiners in line with the new National 5 syllabus and assessment outlines. This is not SQA material but has been devised to provide further guidance for National 5 examinations.

Hodder Gibson is grateful to the copyright holders, as credited below, for permission to use their material. Every effort has been made to trace the copyright holders and to obtain their permission for the use of copyright material. Hodder Gibson will be happy to receive information allowing us to rectify any error or omission in future editions.

Permission has been sought from all relevant copyright holders and Hodder Gibson is grateful for the use of the following:

An extract adapted from 'Potassium Permanganate' by Simon Cotton from 'Education in Chemistry', Nov 2009. Reproduced by permission of Royal Society of Chemistry (2017 SQP Section 2 page 14);
An extract adapted from 'The bizarre world of high pressure chemistry' from 'The Catalyst', Volume 27, Number 1, October 2016, published by the Gatsby Science Enhancement Programme © 2016 STEM Learning Ltd. (2018 Section 2 page 28);
An extract adapted from 'Soundbite Molecules: cold killer – antifreeze' by Dr Simon Cotton from 'Education in Chemistry', May 2008. Reproduced by permission of Royal Society of Chemistry & Dr Simon Cotton (2019 Section 2 page 8);
An extract adapted from 'Beryllium' by Simon Cotton from 'Education in Chemistry', Nov 2015. Reproduced by permission of Royal Society of Chemistry (2019 Section 2 page 20).

Hachette UK's policy is to use papers that are natural, renewable and recyclable products and made from wood grown in well-managed forests and other controlled sources. The logging and manufacturing processes are expected to conform to the environmental regulations of the country of origin.

Orders: please contact Bookpoint Ltd, 130 Park Drive, Milton Park, Abingdon, Oxon OX14 4SE. Telephone: (44) 01235 827827. Fax: (44) 01235 400454. Lines are open 9.00–5.00, Monday to Friday, with a 24-hour message answering service. Visit our website at www.hoddereducation.co.uk. If you have queries or questions that aren't about an order, you can contact us at hoddergibson@hodder.co.uk

This collection first published in 2019 by
Hodder Gibson, an imprint of Hodder Education,
An Hachette UK Company
211 St Vincent Street
Glasgow G2 5QY

Typeset by Aptara, Inc.

Printed in the UK

A catalogue record for this title is available from the British Library

ISBN: 978-1-5104-7816-9

2 1

2020 2019

SCOTLAND EXCEL

We are an approved supplier on the Scotland Excel framework.

Schools can find us on their procurement system as:

Hodder & Stoughton Limited t/a Hodder Gibson.

MIX
Paper from responsible sources
FSC
www.fsc.org
FSC™ C104740

Introduction

National 5 Chemistry

This book of SQA past papers contains the question papers used in the 2018 and 2019 exams (with answers at the back of the book). A specimen question paper reflecting the content and duration of the exam in 2018 is also included.

All of the question papers included in the book provide excellent representative exam practice for the final exams. Using these papers as part of your revision will help you to develop the vital skills and techniques needed for the exam, and will help you identify any knowledge gaps you may have.

It is always a very good idea to refer to SQA's website for the most up-to-date course specification documents. These are available at www.sqa.org.uk/sqa/47428

The course

To achieve a pass in National 5 Chemistry, there are two main components you must complete.

Component 1 – Assignment

You are required to submit an assignment that is worth 20% (20 marks) of your final grade. This assignment will be based on research but it is important that you are aware that practical/experimental/field work is also a mandatory feature of your assignment so has to be included. This assignment requires you to apply skills, knowledge and understanding to investigate a relevant topic in chemistry and its effect on the environment and/or society. Your school or college will provide you with a Candidate's Guide for this assignment, which has been produced by the SQA. This guide gives information on what is required to complete the report and gain as many marks as possible. When you have completed your research and collected your experimental data, you will be allocated a maximum of 1 hour 30 minutes to complete your report in school.

Your assignment report will be marked by the SQA.

Component 2 – The question paper

The question paper will assess breadth and depth of knowledge and understanding from across the whole course. It is worth 80% of your final grade and is marked out of 100 possible marks. The question paper will require you to:

- make statements, provide explanations, and describe information to demonstrate knowledge and understanding.
- apply knowledge and understanding to new situations to solve problems.
- plan and design experiments.

- present information in various forms such as graphs, tables, etc.
- perform calculations based on information given.
- give predictions or make generalisations based on information given.
- draw conclusions based on information given.
- suggest improvement to experiments to improve the accuracy of results obtained or to improve safety.

To achieve a "C" grade in National 5 Chemistry you must achieve about 50% of the 120 marks available when the two components, i.e. the question paper and the assignment are combined. For a "B" you will need 60%, while for an "A" grade you must ensure that you gain as many of the available marks as possible, and at least 70%.

Each SQA Past Paper consists of two sections (a marking scheme for each section is provided at the end of this book):

- Section 1 will contain objective questions (multiple choice) and will have 25 marks.
- Section 2 will contain restricted and extended-response questions and will have 75 marks.

Each SQA Past Paper contains a variety of questions including some that require:

- demonstration and application of knowledge, and understanding of the mandatory content of the course from across the three areas of the course: chemical changes and structure; nature's chemistry; and chemistry in society.
- application of scientific inquiry skills.

How to use this book

This book can be used in two ways:

1. You can complete an entire paper under exam conditions, without the use of books or notes, and then mark the papers using the marking scheme provided. This method gives you a clear indication of the level you are working at and should highlight the content areas that you need to work on before attempting the next paper. This method also allows you to see your progress as you complete each paper.

2. You can complete a paper using your notes and books. Try the question first and then refer to your notes if you are unable to answer the question. This is a form of studying and by doing this you will cover all the areas of content that you are weakest in. You should notice that you are referring to your notes less with each paper completed.

Try to practise as many questions as possible. This will get you used to the language used in the papers and ultimately improve your chances of success.

Some hints and tips

Below is a list of hints and tips that will help you to achieve your full potential in the National 5 exam.

- Ensure that you **read each question carefully**. Scanning the question and missing the main points results in mistakes being made. Some students highlight the main points of a question with a highlighter pen to ensure that they don't miss anything out.

- Open ended questions include the statement **"Using your knowledge of chemistry"**. These questions provide you with an opportunity to show off your chemistry knowledge. To obtain the three marks on offer for these questions, you must demonstrate a good understanding of the chemistry involved and provide a logically correct answer to the question posed.

- When doing calculations, ensure that you **show all of your working**. If you make a simple arithmetical mistake you may still be awarded some of the marks, but only if your working is laid out clearly so that the examiner can see where you went wrong and what you did correctly. Just giving the answers is very risky so you should always show your working.

- **Attempt all questions.** Giving no answer at all means that you will definitely not gain any marks.

- When you are required to read a passage to answer a question, ensure that you **read it carefully** as the information you require is contained within it. It may not be obvious at first, but the answers will be contained within the passage.

- If you are asked to "explain" in a question, then you must **explain your answer fully**. For example, if you are asked to explain how a covalent bond holds atoms together then you cannot simply say:

  "A covalent bond is a shared pair of electrons between atoms in a non-metal."

This answer tells the examiner what a covalent bond is, but does not explain how it holds the atoms together. To gain the marks, an answer similar to this should be written:

  "A covalent bond is a shared pair of electrons between atoms in a non-metal. The shared electrons are attracted to the nuclei of both atoms, which creates a tug-of-war effect, creating the covalent bond."

- You may be required to draw one graph in each exam. To obtain all the marks, ensure that the graphs have **labels**, **units**, **points plotted correctly** and a line of "best fit" drawn between the points.

- Use your **data booklet** when you are asked to write formulas, ionic formulas, formula mass, etc. You have the data booklet in front of you so use it to double check the numbers you require.

- Work on your **timing**. The multiple-choice section (Section 1) should take approximately 40 minutes. Attempt to answer the multiple-choice questions before you look at the four possible answers, as this will improve your confidence. Use scrap paper when required to scribble down structural formulae, calculations, chemical formulae, etc., as this will reduce your chances of making errors. If you are finding the question difficult, try to eliminate the obviously wrong answers to increase your chances.

- When asked to **predict or estimate** based on information from a graph or a table, then take your time to look for patterns. For example, if asked to predict a boiling point, try to establish if there is a regular change in boiling point and use that regular pattern to establish the unknown boiling point.

- When drawing a **diagram** of an experiment ask yourself the question, "Would this work if I set it up exactly like this in the lab?" Ensure that the method you have drawn would produce the desired results *safely*. If, for example, you are heating a flammable reactant such as alcohol then you will not gain the marks if you heat it with a Bunsen burner in your diagram; a water bath would be much safer! Make sure your diagram is labelled clearly.

Good luck!

Remember that the rewards for passing National 5 Chemistry are well worth it! Your pass will help you get the future you want for yourself. In the exam, be confident in your own ability. If you're not sure how to answer a question, trust your instincts and just give it a go anyway. Keep calm and don't panic! GOOD LUCK!

Study Skills – what you need to know to pass exams!

General exam revision: 20 top tips

When preparing for exams, it is easy to feel unsure of where to start or how to revise. This guide to general exam revision provides a good starting place, and, as these are very general tips, they can be applied to all your exams.

1. Start revising in good time.

Don't leave revision until the last minute – this will make you panic and it will be difficult to learn. Make a revision timetable that counts down the weeks to go.

2. Work to a study plan.

Set up sessions of work spread through the weeks ahead. Make sure each session has a focus and a clear purpose. What will you study, when and why? Be realistic about what you can achieve in each session, and don't be afraid to adjust your plans as needed.

3. Make sure you know exactly when your exams are.

Get your exam dates from the SQA website and use the timetable builder tool to create your own exam schedule. You will also get a personalised timetable from your school, but this might not be until close to the exam period.

4. Make sure that you know the topics that make up each course.

Studying is easier if material is in manageable chunks – why not use the SQA topic headings or create your own from your class notes? Ask your teacher for help on this if you are not sure.

5. Break the chunks up into even smaller bits.

The small chunks should be easier to cope with. Remember that they fit together to make larger ideas. Even the process of chunking down will help!

6. Ask yourself these key questions for each course:

- Are all topics compulsory or are there choices?
- Which topics seem to come up time and time again?
- Which topics are your strongest and which are your weakest?

Use your answers to these questions to work out how much time you will need to spend revising each topic.

7. Make sure you know what to expect in the exam.

The subject-specific introduction to this book will help with this. Make sure you can answer these questions:

- How is the paper structured?
- How much time is there for each part of the exam?
- What types of question are involved? These will vary depending on the subject so read the subject-specific section carefully.

8. Past papers are a vital revision tool!

Use past papers to support your revision wherever possible. This book contains the answers and mark schemes too – refer to these carefully when checking your work. Using the mark scheme is useful; even if you don't manage to get all the marks available first time when you first practise, it helps you identify how to extend and develop your answers to get more marks next time – and of course, in the real exam.

9. Use study methods that work well for you.

People study and learn in different ways. Reading and looking at diagrams suits some students. Others prefer to listen and hear material – what about reading out loud or getting a friend or family member to do this for you? You could also record and play back material.

10. There are three tried and tested ways to make material stick in your long-term memory:

- Practising – e.g. rehearsal, repeating
- Organising – e.g. making drawings, lists, diagrams, tables, memory aids
- Elaborating – e.g. incorporating the material into a story or an imagined journey

11. Learn actively.

Most people prefer to learn actively – for example, making notes, highlighting, redrawing and redrafting, making up memory aids, or writing past paper answers. A good way to stay engaged and inspired is to mix and match these methods – find the combination that best suits you. This is likely to vary depending on the topic or subject.

12. Be an expert.

Be sure to have a few areas in which you feel you are an expert. This often works because at least some of them will come up, which can boost confidence.

13. Try some visual methods.

Use symbols, diagrams, charts, flashcards, post-it notes etc. Don't forget – the brain takes in chunked images more easily than loads of text.

14. Remember – practice makes perfect.

Work on difficult areas again and again. Look and read – then test yourself. You cannot do this too much.

15. Try past papers against the clock.

Practise writing answers in a set time. This is a good habit from the start but is especially important when you get closer to exam time.

16. Collaborate with friends.

Test each other and talk about the material – this can really help. Two brains are better than one! It is amazing how talking about a problem can help you solve it.

17. Know your weaknesses.

Ask your teacher for help to identify what you don't know. Try to do this as early as possible. If you are having trouble, it is probably with a difficult topic, so your teacher will already be aware of this – most students will find it tough.

18. Have your materials organised and ready.

Know what is needed for each exam:

- Do you need a calculator or a ruler?
- Should you have pencils as well as pens?
- Will you need water or paper tissues?

19. Make full use of school resources.

Find out what support is on offer:

- Are there study classes available?
- When is the library open?
- When is the best time to ask for extra help?
- Can you borrow textbooks, study guides, past papers, etc.?
- Is school open for Easter revision?

20. Keep fit and healthy!

Try to stick to a routine as much as possible, including with sleep. If you are tired, sluggish or dehydrated, it is difficult to see how concentration is even possible. Combine study with relaxation, drink plenty of water, eat sensibly, and get fresh air and exercise – all these things will help more than you could imagine. Good luck!

NATIONAL 5

2017 Specimen
Question Paper

National
Qualifications
SPECIMEN ONLY

S813/75/02

Chemistry
Section 1—Questions

Date — Not applicable

Duration — 2 hours 30 minutes

Instructions for completion of Section 1 are given on *Page two* of your question and answer booklet S813/75/01.

Record your answers on the answer grid on *Page three* of your question and answer booklet.

You may refer to the Chemistry Data Booklet for National 5.

Before leaving the examination room you must give your question and answer booklet to the Invigilator; if you do not, you may lose all the marks for this paper.

SECTION 1 — 25 marks

Attempt ALL questions

1. Which of the following elements usually exists as diatomic molecules?

   A   Helium

   B   Nitrogen

   C   Silicon

   D   Sulfur

2. Which line in the table correctly describes a proton?

	Mass (atomic mass units)	Charge
A	negligible	+1
B	negligible	−1
C	1	+1
D	1	0

3. Ionic compounds conduct electricity when molten because they have

   A   ions that are free to move

   B   delocalised electrons

   C   metal atoms

   D   a lattice structure.

4. A molecule of phosphine is shown below.

   The shape of a molecule of phosphine is

   A   linear

   B   angular

   C   tetrahedral

   D   trigonal pyramidal.

5. The table gives information about some particles.

Identify the particle which is a negative ion.

| | Number of | | |
Particle	protons	neutrons	electrons
A	9	10	10
B	11	12	11
C	15	16	15
D	19	20	18

6. The table shows the colours of some ionic compounds in solution.

Compound	Colour
copper nitrate	blue
copper chromate	green
strontium nitrate	colourless
strontium chromate	yellow

The colour of the chromate ion is

A   blue

B   green

C   colourless

D   yellow.

7. Which of the following statements correctly describes the concentrations of $H^+(aq)$ and $OH^-(aq)$ ions in pure water?

A   The concentrations of $H^+(aq)$ and $OH^-(aq)$ ions are equal.

B   The concentrations of $H^+(aq)$ and $OH^-(aq)$ ions are zero.

C   The concentration of $H^+(aq)$ ions is greater than the concentration of $OH^-(aq)$ ions.

D   The concentration of $H^+(aq)$ ions is less than the concentration of $OH^-(aq)$ ions.

[Turn over

8.

The name of the above compound is

A   2-ethylpropane

B   1,1-dimethylpropane

C   2-methylbutane

D   3-methylbutane.

9.   Which of the following could be the molecular formula for a cycloalkane?

A C_6H_8

B C_6H_{10}

C C_6H_{12}

D C_6H_{14}

10.   In which of the following types of reaction is oxygen a reactant?

A   Combustion

B   Neutralisation

C   Polymerisation

D   Precipitation

11. Molecules in which four different atoms are attached to a carbon atom are said to be chiral.
Which of the following molecules is chiral?

A

B

C

D

12. Three members of the cycloalkene family are

The general formula for the cycloalkene family is

A C_nH_{2n-2}

B C_nH_{2n-4}

C C_nH_{2n}

D C_nH_{2n+2}

[Turn over

13. Which of the following molecules is an isomer of hept-2-ene?

A

```
     H   H   H   H   H
     |   |   |   |   |
 H — C — C — C — C — C — H
     |   |   |   |   |
     H   H   H   |   H
                 |
             H — C — H
                 |
                 H
```

B

```
     H   H   H   H   H   H
     |   |   |   |   |   |
 H — C — C — C — C — C — C — H
     |   |   |   |   |   |
     H   H   H   |   H   H
                 |
             H — C — H
                 |
                 H
```

C

```
                 H
                 |
             H — C — H
                 |
     H   H   H   H   |   H
     |   |   |   |   |   |
 H — C — C — C — C — C = C — H
     |   |   |   |
     H   H   H   H
```

D

```
     H   H   H   H   H   H   H
     |   |   |   |   |   |   |
 H — C — C — C — C — C = C — C — H
     |   |   |   |           |
     H   H   H   H           H
```

14. A student tested some compounds. The results are given in the table.

Compound	pH of aqueous solution	Effect on bromine solution
CH_3CH_2COOH (propanoic acid structure shown)	4	no effect
$CH_2=CHCOOH$ (propenoic acid structure shown)	4	decolourised
$CH_3CH_2CH_2OH$ (propan-1-ol structure shown)	7	no effect
$CH_2=CHCH_2OH$ (prop-2-en-1-ol structure shown)	7	decolourised

Which line in the table below shows the correct results for the following compound?

$CH_3-CH=CH-CH_2-CH_2-OH$ (structure shown)

	pH of aqueous solution	Effect on bromine solution
A	4	decolourised
B	7	decolourised
C	4	no effect
D	7	no effect

[Turn over

15. Which of the following diagrams could be used to represent the structure of a metal?

A

B

C

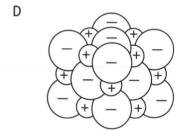

D

16. Which of the following substances does **not** produce water when it reacts with dilute acid?

A   Sodium hydroxide

B   Magnesium

C   Copper oxide

D   Ammonia solution

17. Which of the following metals can be extracted from its oxide by heat alone?

A   Aluminium

B   Zinc

C   Gold

D   Iron

18.

In the cell shown above, electrons flow through

A   the solution from tin to zinc

B   the solution from zinc to tin

C   the connecting wire from tin to zinc

D   the connecting wire from zinc to tin.

19. Four cells were made by joining silver to copper, iron, tin and zinc.

The voltages for the four cells are shown in the table.

Which cell contained silver joined to copper?

You may wish to use the data booklet to help you.

Cell	Voltage (V)
A	1·6
B	1·2
C	0·9
D	0·5

[Turn over

20. The ion-electron equation for the oxidation and reduction steps in the reaction between magnesium and silver(I) ions are:

$$Mg \rightarrow Mg^{2+} + 2e^-$$

$$Ag^+ + e^- \rightarrow Ag$$

The overall redox equation is

A $Mg + 2Ag^+ \rightarrow Mg^{2+} + 2Ag$

B $Mg + Ag^+ \rightarrow Mg^{2+} + Ag$

C $Mg + Ag^+ + e^- \rightarrow Mg^{2+} + Ag + 2e^-$

D $Mg + 2Ag \rightarrow Mg^{2+} + 2Ag^+$.

21. The structure below shows a section of an addition polymer.

Which of the following molecules is used to make this polymer?

A

B

C

D

22. Hydrogen gas

    A   burns with a pop

    B   relights a glowing splint

    C   turns damp pH paper red

    D   turns limewater cloudy.

23. What is the charge on an iron ion in $Fe_2(SO_4)_3$?

    A   3–

    B   3+

    C   2–

    D   2+

24. Sodium sulfate solution reacts with barium chloride solution.

    $Na_2SO_4(aq) + BaCl_2(aq) \rightarrow BaSO_4(s) + 2NaCl(aq)$

    The spectator ions present in this reaction are

    A Ba^{2+} and Cl^-

    B Ba^{2+} and SO_4^{2-}

    C Na^+ and Cl^-

    D Na^+ and SO_4^{2-}

[Turn over

25. But-1-ene is a colourless, insoluble gas which is more dense than air but less dense than water.

Which of the following diagrams shows the most appropriate apparatus for collecting and measuring the volume of but-1-ene?

A — measuring cylinder

B — measuring cylinder

C — measuring cylinder
tub
water

D — measuring cylinder
water

[END OF SECTION 1. NOW ATTEMPT THE QUESTIONS IN SECTION 2 OF YOUR QUESTION AND ANSWER BOOKLET]

N5

National
Qualifications
SPECIMEN ONLY

Mark

S813/75/01

Chemistry
Section 1—Answer Grid
and Section 2

Date — Not applicable

Duration — 2 hours 30 minutes

Fill in these boxes and read what is printed below.

Full name of centre

Town

Forename(s)

Surname

Number of seat

Date of birth

Day   Month   Year

Scottish candidate number

Total marks — 100

SECTION 1 — 25 marks

Attempt ALL questions.

Instructions for the completion of Section 1 are given on *Page two*.

SECTION 2 — 75 marks

Attempt ALL questions.

You may refer to the Chemistry Data Booklet for National 5.

Write your answers clearly in the spaces provided in this booklet. Additional space for answers and rough work is provided at the end of this booklet. If you use this space you must clearly identify the question number you are attempting. Any rough work must be written in this booklet. Score through your rough work when you have written your final copy.

Use **blue** or **black** ink.

Before leaving the examination room you must give this booklet to the Invigilator; if you do not, you may lose all the marks for this paper.

SECTION 1 — 25 marks

The questions for Section 1 are contained in the question paper S813/75/02.

Read these and record your answers on the answer grid on *Page three* opposite.

Use **blue** or **black** ink. Do NOT use gel pens or pencil.

1. The answer to each question is **either** A, B, C, or D. Decide what your answer is, then fill in the appropriate bubble (see sample question below).

2. There is **only one correct** answer to each question.

3. Any rough working should be done on the additional space for answers and rough work at the end of this booklet.

Sample Question

To show that the ink in a ball-pen consists of a mixture of dyes, the method of separation would be

    A   fractional distillation

    B   chromatography

    C   fractional crystallisation

    D   filtration.

The correct answer is **B** — chromatography. The answer **B** bubble has been clearly filled in (see below).

Changing an answer

If you decide to change your answer, cancel your first answer by putting a cross through it (see below) and fill in the answer you want. The answer below has been changed to **D**.

If you then decide to change back to an answer you have already scored out, put a tick (✓) to the **right** of the answer you want, as shown below:

SECTION 1 — Answer Grid

	A	B	C	D
1	○	○	○	○
2	○	○	○	○
3	○	○	○	○
4	○	○	○	○
5	○	○	○	○
6	○	○	○	○
7	○	○	○	○
8	○	○	○	○
9	○	○	○	○
10	○	○	○	○
11	○	○	○	○
12	○	○	○	○
13	○	○	○	○
14	○	○	○	○
15	○	○	○	○
16	○	○	○	○
17	○	○	○	○
18	○	○	○	○
19	○	○	○	○
20	○	○	○	○
21	○	○	○	○
22	○	○	○	○
23	○	○	○	○
24	○	○	○	○
25	○	○	○	○

[BLANK PAGE]

DO NOT WRITE ON THIS PAGE

MARKS | DO NOT WRITE IN THIS MARGIN

SECTION 2 — 75 marks

Attempt ALL questions

1. Graphs can be used to show the change in the rate of a reaction as the reaction proceeds.

    The graph shows the volume of gas produced in an experiment over a period of time.

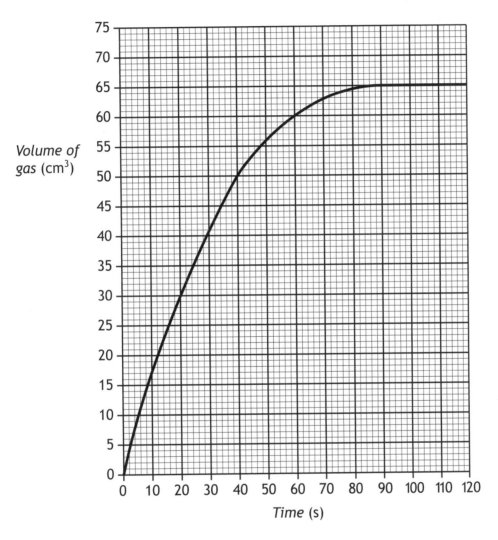

    (a) State the time, in seconds, at which the reaction stopped.   1

[Turn over

MARKS | DO NOT WRITE IN THIS MARGIN

1. **(continued)**

(b) Calculate the average rate of reaction for the first 20 seconds.

Your answer must include the appropriate unit.

Show your working clearly.

3

(c) The graph shows that the rate of reaction decreases as the reaction proceeds.

Suggest a reason for this decrease.

1

MARKS | DO NOT WRITE IN THIS MARGIN

2. The group 7 element bromine was discovered by Balard in 1826.

   Bromine gets its name from the Greek "bromos" meaning stench.

   A sample of bromine consists of a mixture of two isotopes, $^{79}_{35}Br$ and $^{81}_{35}Br$.

   (a) State what is meant by the term isotope.   **1**

   (b) Complete the table for $^{79}_{35}Br$.   **1**

Isotope	Number of protons	Number of neutrons
$^{79}_{35}Br$		

   (c) The sample of bromine has an average atomic mass of 80.

   Suggest what this indicates about the amount of each isotope in this sample.   **1**

   [Turn over

MARKS | DO NOT WRITE IN THIS MARGIN

2.   **(continued)**

(d) In 1825 bromine had been isolated from sea water by Liebig who mistakenly thought it was a compound of iodine and chlorine.

Using your knowledge of chemistry, comment on why Liebig might have made this mistake.

3

MARKS | DO NOT WRITE IN THIS MARGIN

3. Antacid tablets are used to treat indigestion which is caused by excess acid in the stomach.

Different brands of tablets contain different active ingredients.

Name of active ingredient	magnesium carbonate	calcium carbonate	magnesium hydroxide	aluminium hydroxide
Reaction with acid	fizzes	fizzes	does not fizz	does not fizz
Cost per gram (pence)	16	11	7·5	22
Mass of solid needed to neutralise 20 cm^3 of acid (g)	0·7	1·2	0·6	0·4
Cost of neutralising 20 cm^3 of acid (pence)		13·2	4·5	8·8

(a) Write the formula, showing the charge on each ion, for aluminium hydroxide.   1

(b) (i) Complete the table to show the cost of using magnesium carbonate to neutralise 20 cm^3 of acid.   1

(ii) Using information from the table, state which **one** of the four active ingredients **you** would use to neutralise excess stomach acid.

Explain your choice.   1

[Turn over

MARKS | DO NOT WRITE IN THIS MARGIN

4. Sulfur dioxide gas is produced when fossil fuels containing sulfur are burned.

(a) When sulfur dioxide dissolves in water in the atmosphere "acid rain" is produced.

Circle the correct phrase to complete the sentence.   **1**

Compared with pure water, acid rain contains $\left\{\begin{array}{l}\text{a higher}\\\text{a lower}\\\text{the same}\end{array}\right\}$ concentration of hydrogen ions.

(b) The table gives information about the solubility of sulfur dioxide.

Temperature (°C)	18	24	30	36	42	48
Solubility (g/100 cm³)	11·2	9·2	7·8	6·5	5·5	4·7

(i) Draw a graph of solubility against temperature.

Use appropriate scales to fill most of the graph paper.   **4**

(Additional graph paper, if required, can be found on *Page twenty-eight*.)

MARKS | DO NOT WRITE IN THIS MARGIN

4. **(b)** **(continued)**

(ii) Estimate the solubility of sulfur dioxide, in g/100 cm^3, at 21 °C.   1

[Turn over

MARKS | DO NOT WRITE IN THIS MARGIN

5. A student investigated the reaction of carbonates with dilute hydrochloric acid.

   (a) In one reaction lithium carbonate reacted with dilute hydrochloric acid.
   The equation for the reaction is:

   $$Li_2CO_3(s) + HCl(aq) \rightarrow LiCl(aq) + CO_2(g) + H_2O(\ell)$$

      (i) Balance this equation.   1

      (ii) Identify the salt produced in this reaction.   1

   (b) In another reaction 1·0 g of calcium carbonate reacted with excess dilute hydrochloric acid.

   $$CaCO_3(s) + 2HCl(aq) \rightarrow CaCl_2(aq) + CO_2(g) + H_2O(\ell)$$

      (i) Calculate the mass, in grams, of carbon dioxide produced.   3

      Show your working clearly.

MARKS | DO NOT WRITE IN THIS MARGIN

5. (b) (continued)

(ii) The student considered two methods to confirm the mass of carbon dioxide gas produced in this reaction.

Method A   *Method B*

Method A	Method B
1. Add the acid from the measuring cylinder to the calcium carbonate in the flask.	1. Weigh the flask with the calcium carbonate and the acid in the measuring cylinder together.
2. Weigh the flask and contents.	2. Add the acid from the measuring cylinder to the calcium carbonate in the flask and replace the empty measuring cylinder on the balance.
3. Leave until no more bubbles are produced.	3. Leave until no more bubbles are produced.
4. Reweigh the flask and contents.	4. Reweigh the flask, contents and the empty measuring cylinder together.

Explain which method would give a more reliable estimate of the mass of carbon dioxide produced during the reaction.   2

[Turn over

MARKS | DO NOT WRITE IN THIS MARGIN

6. Read the passage below and answer the questions that follow.

Potassium Permanganate ($KMnO_4$)

Potassium permanganate's strong oxidising properties make it an effective disinfectant. Complaints such as athlete's foot and some fungal infections are treated by bathing the affected area in $KMnO_4$ solution.

In warm climates vegetables are washed in $KMnO_4$ to kill bacteria such as *E. coli*. Chemists use $KMnO_4$ in the manufacture of saccharin and benzoic acid.

Baeyer's reagent is an alkaline solution of $KMnO_4$ and is used to detect unsaturated organic compounds. The reaction of $KMnO_4$ with alkenes is also used to extend the shelf life of fruit. Ripening fruit releases ethene gas which causes other fruit to ripen. Shipping containers are fitted with gas scrubbers that use alumina or zeolite impregnated with $KMnO_4$ to stop the fruit ripening too quickly.

$$C_2H_4 + 4KMnO_4 \rightarrow 4MnO_2 + 4KOH + 2CO_2$$

Adapted from an article by Simon Cotton on "Soundbite molecules" in "Education in Chemistry" November 2009.

(a) Suggest an experimental test, including the result, to show that potassium is present in potassium permanganate.   1

You may wish to use the data booklet to help you.

(b) Suggest a pH for Baeyer's reagent.   1

(c) Name the gas removed by the scrubbers.   1

(d) Name a chemical mentioned in the passage which contains the following functional group.   1

(e) Zeolite is a substance that contains aluminium silicate.

Name the elements present in aluminium silicate.   1

MARKS | DO NOT WRITE IN THIS MARGIN

7. In the 2012 London Olympics, alkanes were used as fuels for the Olympic flame.

   (a) The torches that carried the Olympic flame across Britain burned a mixture of propane and butane.

   Propane and butane are members of the same homologous series.

   State what is meant by the term homologous series. **1**

   (b) Natural gas, which is mainly methane, was used to fuel the flame in the Olympic cauldron.

       (i) Draw a diagram to show how **all** the outer electrons are arranged in a molecule of methane, CH_4. **1**

       (ii) Methane is a covalent molecular substance. It has a low boiling point and is a gas at room temperature.

       Explain why methane is a gas at room temperature. **2**

[Turn over

MARKS | DO NOT WRITE IN THIS MARGIN

8. Car manufacturers have developed vehicles that use ethanol as fuel.

(a) The structure of ethanol is shown below.

Name the functional group circled in the diagram.   **1**

(b) Name the two substances produced when ethanol burns in a plentiful supply of oxygen.   **1**

(c) Ethanol can be produced from ethene as shown.

ethene   ethanol

(i) Name the **type** of chemical reaction taking place.   **1**

MARKS | DO NOT WRITE IN THIS MARGIN

8. (c) (continued)

(ii) Draw a structural formula for a product of the following reaction.　　1

(d) Ethanol can be used to produce ethanoic acid.

(i) Draw a structural formula for ethanoic acid.　　1

(ii) Name the family of compounds to which ethanoic acid belongs.　　1

[Turn over

MARKS | DO NOT WRITE IN THIS MARGIN

9. Alkanes burn, releasing heat energy.

(a) State the term used to describe all chemical reactions that release heat energy.

1

(b) A student investigated the amount of energy released when an alkane burns using the apparatus shown.

The student recorded the following data.

Mass of alkane burned	1 g
Volume of water	200 cm^3
Initial temperature of water	15 °C
Final temperature of water	55 °C

(i) Calculate the energy released, in kJ.

Show your working clearly.

3

MARKS | DO NOT WRITE IN THIS MARGIN

9. (b) (continued)

   (ii) Suggest **one** improvement to the student's investigation.   1

(c) The table gives information about the amount of energy released when one mole of some alkanes are burned.

Name of alkane	Energy released when one mole of alkane is burned (kJ)
methane	891
ethane	1561
propane	2219
butane	2878

   (i) Write a statement linking the amount of energy released to the number of carbon atoms in the alkane molecule.   1

   (ii) Predict the amount of heat released, in kJ, when one mole of pentane is burned.   1

[Turn over

MARKS | DO NOT WRITE IN THIS MARGIN

10. Essential oils can be extracted from plants and used in perfumes and food flavourings.

(a) Essential oils contain compounds made up of a number of isoprene molecules joined together.

The shortened structural formula for isoprene is $CH_2C(CH_3)CHCH_2$.

Draw the **full** structural formula for isoprene.

1

(b) Essential oils can be extracted from the zest of lemons in the laboratory by steam distillation.

The process involves heating up water in a boiling tube until it boils. The steam produced then passes over the lemon zest which is separated from the water by glass wool. As the steam passes over the lemon zest it carries essential oils into the delivery tube. The condensed liquids (essential oils and water) are collected in a test tube placed in a cold water bath.

Complete the diagram to show the apparatus needed to collect the essential oils.

1

(An additional diagram, if required, can be found on *Page twenty-nine*.)

lemon zest

water

glass wool

heat

MARKS | DO NOT WRITE IN THIS MARGIN

10.   (continued)

(c)   Limonene, $C_{10}H_{16}$, is a compound found in lemon zest.

Write the molecular formula for the product formed when limonene reacts completely with bromine solution.

1

[Turn over

MARKS | DO NOT WRITE IN THIS MARGIN

11. Metals can be extracted from metal compounds by electrolysis.

(a) During electrolysis, metal ions are changed to metal atoms.

Name this type of chemical reaction.   1

(b) A student set up the following experiment to electrolyse copper(II) chloride solution.

(i) Name the type of power supply that **must** be used to electrolyse the solution.   1

(ii) Complete the table by adding the charge on each electrode.   1

Observation at the _____ electrode	Observation at the _____ electrode
brown solid formed	bubbles of gas

MARKS | DO NOT WRITE IN THIS MARGIN

12. Urea, H_2NCONH_2, can be used as a fertiliser.

(a) Calculate the percentage of nitrogen in urea.

3

(b) Other nitrogen based fertilisers can be produced from ammonia.
Ammonia is produced in an industrial process using a catalyst.

$$N_2(g) \ + \ 3H_2(g) \ \rightleftharpoons \ 2NH_3(g)$$

(i) Name the industrial process that produces ammonia.

1

(ii) Suggest why a catalyst may be used in an industrial process.

1

(c) In another industrial process, ammonia is used to produce nitric acid.
Name the catalyst used in this process.

1

[Turn over

MARKS | DO NOT WRITE IN THIS MARGIN

13. Vitamin C is found in fruits and vegetables.

0·0050 mol l⁻¹ iodine solution

flask

25 cm³ orange juice containing vitamin C + indicator

Using iodine solution, a student carried out experiments to determine the concentration of vitamin C in orange juice.

The results of the experiments are shown.

Experiment	Initial volume of iodine solution (cm³)	Final volume of iodine solution (cm³)	Volume of iodine solution added (cm³)
1	1·2	18·0	16·8
2	18·0	33·9	15·9
3	0·5	16·6	16·1

(a) (i) Name the piece of apparatus used to measure the volume of iodine solution added to the orange juice.   1

(ii) Calculate the average volume, in cm³, of iodine solution that should be used in calculating the concentration of vitamin C.   1

Show your working clearly.

(b) Name the experimental method, carried out by the student, to accurately determine the concentration of vitamin C in the orange juice.   1

14. In medicine, technetium-99m is injected into the body to detect damage to heart tissue.

It is a gamma-emitting radioisotope with a half-life of 6 hours.

(a) A sample of technetium-99m has a mass of 2 g.

Calculate the mass, in grams, of technetium-99m that would be left after 12 hours.

2

Show your working clearly.

(b) Suggest one reason why technetium-99m can be used safely in this way.

1

(c) Technetium-99m is formed when molybdenum-99 decays.

The decay equation is:

$$^{99}_{42}Mo \rightarrow ^{99}_{43}Tc + X$$

Identify **X**.

1

[Turn over

MARKS | DO NOT WRITE IN THIS MARGIN

15. The concentration of chloride ions in water affects the ability of some plants to grow.

A student investigated the concentration of chloride ions in the water at various points along the river Tay.

The concentration of chloride ions in water can be determined by reacting the chloride ions with silver ions.

$$Ag^+(aq) \ + \ Cl^-(aq) \ \rightarrow \ AgCl(s)$$

A $20\,cm^3$ water sample gave a precipitate of silver chloride with a mass of $1 \cdot 435\,g$.

(a) Calculate the number of moles of silver chloride, AgCl, present in this sample.

2

(b) Using your answer to part (a), calculate the concentration, in $mol\,l^{-1}$, of chloride ions in this sample.

2

16. Nitrogen, phosphorus and potassium are elements essential for plant growth.

A student was asked to prepare a dry sample of a compound which contained **two** of these elements.

The student was given access to laboratory equipment and the following chemicals.

Chemical	Formula
ammonium hydroxide	NH_4OH
magnesium nitrate	$Mg(NO_3)_2$
nitric acid	HNO_3
phosphoric acid	H_3PO_4
potassium carbonate	K_2CO_3
potassium hydroxide	KOH
sodium hydroxide	$NaOH$
sulfuric acid	H_2SO_4
water	H_2O

Using your knowledge of chemistry, comment on how the student could prepare their dry sample.

3

[END OF SPECIMEN QUESTION PAPER]

MARKS | DO NOT WRITE IN THIS MARGIN

ADDITIONAL SPACE FOR ANSWERS

Additional graph paper for Question 4 (b) (i)

MARKS | DO NOT WRITE IN THIS MARGIN

ADDITIONAL SPACE FOR ANSWERS

Additional diagram for Question 10 (b)

ADDITIONAL SPACE FOR ANSWERS AND ROUGH WORK

ADDITIONAL SPACE FOR ANSWERS AND ROUGH WORK

NATIONAL 5

2018

National Qualifications 2018

X813/75/02

Chemistry
Section 1—Questions

MONDAY, 21 MAY

1:00 PM — 3:30 PM

Instructions for completion of Section 1 are given on *Page two* of your question and answer booklet X813/75/01.

Record your answers on the answer grid on *Page three* of your question and answer booklet.

You may refer to the Chemistry Data Booklet for National 5.

Before leaving the examination room you must give your question and answer booklet to the Invigilator; if you do not, you may lose all the marks for this paper.

SECTION 1 — 25 marks
Attempt ALL questions

1. Which of the following changes would **not** speed up a chemical reaction?

    A   Increasing the particle size

    B   Increasing the temperature

    C   Increasing the concentration

    D   Addition of a catalyst

2. Which line in the table identifies the correct location of a proton and an electron in an atom?

	Proton	Electron
A	inside the nucleus	inside the nucleus
B	inside the nucleus	outside the nucleus
C	outside the nucleus	outside the nucleus
D	outside the nucleus	inside the nucleus

3. Which of the following elements does **not** exist as diatomic molecules?

    A   Oxygen

    B   Helium

    C   Bromine

    D   Hydrogen

4. The shapes of some molecules are shown below.

tetrahedral   trigonal   angular   linear
                     pyramidal

The shape of a molecule of hydrogen bromide is likely to be

    A   tetrahedral

    B   trigonal pyramidal

    C   angular

    D   linear.

5. Which of the following elements forms an ion with a single positive charge and an electron arrangement of 2,8?

You may wish to use the data booklet to help you.

A   Sodium

B   Magnesium

C   Fluorine

D   Neon

6. Which line in the table shows the properties of a covalent network compound?

	Melting point (°C)	Boiling point (°C)	Conducts electricity	
			Solid	Liquid
A	−127	−100	no	no
B	795	1410	no	yes
C	30	2204	yes	yes
D	2700	3350	no	no

7. 0·1 mol of sodium hydroxide was dissolved in water and the solution made up to 250 cm³.

What is the concentration, in mol l⁻¹, of the sodium hydroxide solution?

A   0·0004

B   0·025

C   0·4

D   2·5

8. An alkaline solution contains

A   only hydroxide ions

B   more hydroxide ions than hydrogen ions

C   more hydrogen ions than hydroxide ions

D   equal numbers of hydrogen ions and hydroxide ions.

[Turn over

9. A student made some statements about the effect of adding water to an acidic solution.
    Identify the correct statement.

    A   The pH of the solution will remain the same.

    B   The pH of the solution will decrease.

    C   The hydrogen ion concentration will decrease.

    D   The hydrogen ion concentration will increase.

10. The shortened structural formula for a compound is

$$CH_3CH_2CH(CH_3)CH(CH_3)CH_2CH_2CH_3$$

Which of the following is another way of representing this structure?

A
```
        H   CH3  H   CH3  H   H   H
        |   |    |   |    |   |   |
   H — C — C — C — C — C — C — C — H
        |   |    |   |    |   |   |
        H   H    H   H    H   H   H
```

B
```
        H   H   H   CH3  H   H
        |   |   |   |    |   |
   H — C — C — C — C — C — C — H
        |   |   |   |    |   |
        H   H   CH3  H   H   H
```

C
```
        H   H   CH3  H   H   H   H
        |   |   |    |   |   |   |
   H — C — C — C — C — C — C — C — H
        |   |   |    |   |   |   |
        H   H   CH3  H   H   H   H
```

D
```
        H   H   H   CH3  H   H   H
        |   |   |   |    |   |   |
   H — C — C — C — C — C — C — C — H
        |   |   |   |    |   |   |
        H   H   CH3  H   H   H   H
```

11. Identify which of the following is an isomer of

```
     H   H   H   H   H
     |   |   |   |   |
 H—C—C—C—C—C—H
     |   |   |   |   |
     H   |   H   H   H
         H—C—H
             |
             H
```

A
```
          H       H
           \     /
            C
       H   / \   H
        \ /     \ /
     H—C         C—H
     H—C         C—H
        / \     / \
       H   C   H
            / \
           H   H
```

B
```
     H   H   H   H   H
     |   |   |   |   |
 H—C—C—C—C—C—H
     |   |   |   |   |
     H   H   H   |   H
                 H—C—H
                     |
                     H
```

C
```
     H   H   H   H   H
     |   |   |   |   |
 H—C—C—C—C—C—H
     |   |   |   |   |
     H   H   |   H   H
             H—C—H
                 |
                 H
```

D
```
     H   H   H   H
     |   |   |   |
 H—C—C—C—C—H
     |   |   |   |
     H   H   |   H
             H—C—H
                 |
                 H
```

[Turn over

12. Which of the following reactions takes place when an alcohol is formed from an alkene?

    A Hydrogenation

    B Combustion

    C Hydration

    D Reduction

13.

The systematic name for the above compound is

    A pentan-2-ol

    B pentan-4-ol

    C 1-methylbutan-3-ol

    D 4-methylbutan-2-ol.

14. Which of the following alcohols is the least soluble in water?

    A Butan-1-ol

    B Hexan-1-ol

    C Pentan-1-ol

    D Propan-1-ol

15. A student set up an experiment to determine the quantity of energy released when a hydrocarbon burns.

Which of the following diagrams shows the apparatus which would produce the most accurate result?

A

B

C

D

[Turn over

16. The ether, 1-ethoxypropane, can be made by the Williamson reaction.

ethanol 1-chloropropane 1-ethoxypropane

The structural formula for another ether is shown below.

2-ethoxypropane

Which of the following pairs of compounds would react together to produce 2-ethoxypropane?

A

B

C

D

17. Information about the reactions of four different metals, **W**, **X**, **Y** and **Z** is given in the table.

Metal	Reaction with dilute acid	Reaction with water
W	moderate reaction	no reaction
X	fast reaction	slow reaction
Y	slow reaction	no reaction
Z	fast reaction	no reaction

The order of reactivity of the metals, starting with the most reactive is

A　X, Z, W, Y

B　Y, W, Z, X

C　Z, X, W, Y

D　Y, W, X, Z.

18. The ion-electron equations for the oxidation and reduction steps in the reaction between hydrogen and oxygen are given below.

$$H_2(g) \longrightarrow 2H^+(aq) + 2e^-$$

$$2H_2O(\ell) + O_2(g) + 4e^- \longrightarrow 4OH^-(aq)$$

The redox equation for the overall reaction is

A　$H_2(g) + 2H_2O(\ell) + O_2(g) + 4e^- \longrightarrow 2H^+(aq) + 4OH^-(aq) + 2e^-$

B　$2H_2(g) + 2H_2O(\ell) + O_2(g) \longrightarrow 4H^+(aq) + 4OH^-(aq)$

C　$H_2(g) + 2H_2O(\ell) + O_2(g) \longrightarrow 2H^+(aq) + 4OH^-(aq)$

D　$2H_2(g) + 2H_2O(\ell) + O_2(g) + 4e^- \longrightarrow 4H^+(aq) + 4OH^-(aq) + 4e^-$

[Turn over

19. Which of the following metals, when connected to lead in a cell, would produce the highest reading on the voltmeter?

You may wish to use the data booklet to help you.

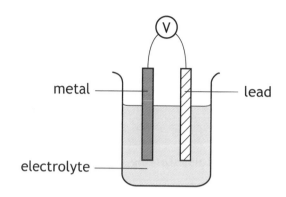

A   Zinc

B   Tin

C   Nickel

D   Lead

20. Which of the following salts would **not** be used as a fertiliser?

A   Ammonium chloride

B   Ammonium phosphate

C   Sodium chloride

D   Sodium phosphate

21. Which metal is used as the catalyst in the industrial manufacture of ammonia?

    A   Nickel

    B   Platinum

    C   Iron

    D   Rhodium

22. The diagram shows the path of two different types of radiation as they pass through an electric field.

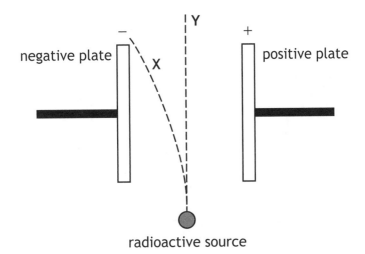

Which line in the table correctly identifies the types of radiation which follow paths X and Y?

	Path X	Path Y
A	alpha	beta
B	beta	alpha
C	beta	gamma
D	alpha	gamma

23. Metallic bonding is a force of attraction between

    A   a shared pair of electrons and two nuclei

    B   negative ions and delocalised electrons

    C   negative ions and positive ions

    D   positive ions and delocalised electrons.

[Turn over

24. $2K^+(aq) + 2I^-(aq) + Pb^{2+}(aq) + 2NO_3^-(aq) \rightarrow PbI_2(s) + 2K^+(aq) + 2NO_3^-(aq)$

The type of reaction represented by this equation is

A   neutralisation

B   precipitation

C   addition

D   redox.

25. A student prepared a sample of copper sulfate crystals by reacting excess copper carbonate with acid.

X   Y   Z

Which line in the table shows the correct order in which this experiment would be carried out?

A   Y, X, Z

B   X, Y, Z

C   Z, Y, X

D   Y, Z, X

[END OF SECTION 1. NOW ATTEMPT THE QUESTIONS IN SECTION 2 OF YOUR QUESTION AND ANSWER BOOKLET]

FOR OFFICIAL USE

National Qualifications 2018

Mark

X813/75/01

Chemistry
Section 1—Answer Grid
and Section 2

MONDAY, 21 MAY

1:00 PM – 3:30 PM

Fill in these boxes and read what is printed below.

Full name of centre

Town

Forename(s)

Surname

Number of seat

Date of birth

Day	Month	Year	Scottish candidate number

Total marks — 100

SECTION 1 — 25 marks

Attempt ALL questions.

Instructions for the completion of Section 1 are given on *Page two*.

SECTION 2 — 75 marks

Attempt ALL questions.

You may refer to the Chemistry Data Booklet for National 5.

Write your answers clearly in the spaces provided in this booklet. Additional space for answers and rough work is provided at the end of this booklet. If you use this space you must clearly identify the question number you are attempting. Any rough work must be written in this booklet. You should score through your rough work when you have written your final copy.

Use **blue** or **black** ink.

Before leaving the examination room you must give this booklet to the Invigilator; if you do not, you may lose all the marks for this paper.

SECTION 1 — 25 marks

The questions for Section 1 are contained in the question paper X813/75/02.

Read these and record your answers on the answer grid on *Page three* opposite.

Use **blue** or **black** ink. Do NOT use gel pens or pencil.

1. The answer to each question is **either** A, B, C or D. Decide what your answer is, then fill in the appropriate bubble (see sample question below).

2. There is **only one correct** answer to each question.

3. Any rough working should be done on the additional space for answers and rough work at the end of this booklet.

Sample question

To show that the ink in a ball-pen consists of a mixture of dyes, the method of separation would be

A   fractional distillation

B   chromatography

C   fractional crystallisation

D   filtration.

The correct answer is **B** — chromatography. The answer **B** bubble has been clearly filled in (see below).

Changing an answer

If you decide to change your answer, cancel your first answer by putting a cross through it (see below) and fill in the answer you want. The answer below has been changed to **D**.

If you then decide to change back to an answer you have already scored out, put a tick (✓) to the **right** of the answer you want, as shown below:

SECTION 1 — Answer Grid

	A	B	C	D
1	○	○	○	○
2	○	○	○	○
3	○	○	○	○
4	○	○	○	○
5	○	○	○	○
6	○	○	○	○
7	○	○	○	○
8	○	○	○	○
9	○	○	○	○
10	○	○	○	○
11	○	○	○	○
12	○	○	○	○
13	○	○	○	○
14	○	○	○	○
15	○	○	○	○
16	○	○	○	○
17	○	○	○	○
18	○	○	○	○
19	○	○	○	○
20	○	○	○	○
21	○	○	○	○
22	○	○	○	○
23	○	○	○	○
24	○	○	○	○
25	○	○	○	○

[BLANK PAGE]

DO NOT WRITE ON THIS PAGE

[Turn over for next question

DO NOT WRITE ON THIS PAGE

MARKS | DO NOT WRITE IN THIS MARGIN

SECTION 2 — 75 marks

Attempt ALL questions

1. A student monitored the rate of reaction between excess calcium carbonate and dilute hydrochloric acid, HCl, using a gas syringe to collect the gas produced.

$75\,cm^3$ of $0\cdot1\,mol\,l^{-1}$ hydrochloric acid

calcium carbonate

(a) Name the gas produced in this reaction.　　1

(b) The student obtained the results shown.

Time (s)	0	10	20	40	50	60	70	80
Volume of gas (cm³)	0	48	62	74	77	79	80	80

    (i) Calculate the average rate of reaction between 20 and 50 seconds.　　3

        Your answer must include the appropriate unit.

        Show your working clearly.

MARKS | DO NOT WRITE IN THIS MARGIN

1. (b) (continued)

(ii) Draw a graph of the student's results.   4

(Additional graph paper, if required, can be found on *Page thirty-three*).

(iii) **Using your graph**, estimate the volume of gas, in cm³, produced at 30 seconds.   1

(c) The student carried out a similar experiment using 75 cm³ of 0·1 mol l⁻¹ sulfuric acid, H_2SO_4(aq).

The total volume of gas collected was 160 cm³.

Explain why there was a greater volume of gas produced.   1

MARKS | DO NOT WRITE IN THIS MARGIN

2. The retractable roof on Centre Court at Wimbledon Tennis Club is made of the polymer poly(tetrafluoroethene), PTFE.

(a) The monomer used to produce PTFE has the following structure.

$$
\begin{array}{cc}
F & F \\
| & | \\
C = C \\
| & | \\
F & F
\end{array}
$$

tetrafluoroethene

(i) Name the type of polymerisation used to produce PTFE.   1

(ii) Draw a section of poly(tetrafluoroethene) showing three monomer units joined together.   1

(b) The roof of the O_2 Arena in London is made from a co-polymer.

A co-polymer is formed when two different monomers polymerise.

The repeating unit of the co-polymer is shown.

$$
\begin{bmatrix}
& H & H & F & F & \\
& | & | & | & | & \\
- & C & - C & - C & - C & - \\
& | & | & | & | & \\
& H & H & F & F &
\end{bmatrix}_n
$$

One of the monomers in this co-polymer is tetrafluoroethene.

Draw the full structural formula for the other monomer.   1

Page eight

MARKS | DO NOT WRITE IN THIS MARGIN

3. Coal is a fuel that contains carbon. Different types of coal contain different percentages of carbon.

Heat content is a measure of how much heat energy is released when coal is burned.

(a) The table gives information about types of coal.

Type of coal	Percentage of carbon	Average heat content ($kJ\,kg^{-1}$)
Anthracite	86—98	32 500
Bituminous	45—85	27 850
Sub-bituminous	35—44	25 550
Lignite	25—34	13 950

Describe how the percentage of carbon in coal affects the average heat content.

1

(b) Iron pyrite, FeS_2, is an impurity found in coal.

Calculate the percentage of iron in iron pyrite.

Show your working clearly.

3

MARKS | DO NOT WRITE IN THIS MARGIN

4. During the FIFA World Cup, referees will spray foam onto the pitch to ensure players stand the correct distance from the ball when a free kick is taken. The foam contains a hydrocarbon mixture of isobutane, butane and propane.

(a) Name the elements present in a hydrocarbon. **1**

(b) The full structural formula for isobutane is

$$
\begin{array}{ccccc}
 & & H & & \\
 & & | & & \\
 & H— & C & —H & \\
 & H & | & H & \\
 & | & | & | & \\
H—& C—& C—& C & —H \\
 & | & | & | & \\
 & H & H & H &
\end{array}
$$

Write the systematic name for isobutane. **1**

(c) Alkanes have different physical properties.

The table gives some information about isobutane and butane.

Alkane	Boiling point (°C)
isobutane	−12
butane	−1

Circle the correct words to complete the sentence. **1**

Compared to isobutane, butane has a higher boiling point

as it contains $\left\{ \begin{array}{c} \text{weaker} \\ \text{stronger} \end{array} \right\}$ $\left\{ \begin{array}{c} \text{covalent bonds} \\ \text{intermolecular forces} \end{array} \right\}$.

MARKS | DO NOT WRITE IN THIS MARGIN

4. **(continued)**

(d) The table shows the boiling points of some alkanes.

Alkane	Boiling point (°C)
pentane	36
hexane	69
heptane	98
octane	126
nonane	

Predict the boiling point, in °C, of nonane, C_9H_{20}.   1

[Turn over

MARKS | DO NOT WRITE IN THIS MARGIN

5. Read the passage and answer the questions that follow.

The Chemistry within Airbags

Airbags, an important safety feature in cars, inflate rapidly on collision. Inside the airbag is a gas generator containing a mixture of sodium azide (NaN_3), potassium nitrate and silicon dioxide.

When a car is involved in a collision, a series of three chemical reactions takes place.

In the first reaction, electrical energy causes sodium azide to decompose producing sodium metal and nitrogen gas. The nitrogen gas that is generated fills the airbag.

In the second reaction, the sodium reacts with potassium nitrate producing more nitrogen gas, sodium oxide and potassium oxide.

In the final reaction, the metal oxides react with silicon dioxide to produce silicate fibres, which are harmless and stable.

This process, from the initial impact of the crash to full inflation of the airbag, takes a fraction of a second.

(a) Name the three chemicals found inside the gas generator before any chemical reactions take place.

1

(b) Name the compound produced in the second reaction which would give a lilac flame colour.

You may wish to use the data booklet to help you.

1

(c) Write the formula for the compound which reacts with the metal oxides in the final reaction.

1

MARKS | DO NOT WRITE IN THIS MARGIN

5. (continued)

(d) The graph below gives information on the volume of nitrogen gas produced by the gas generator.

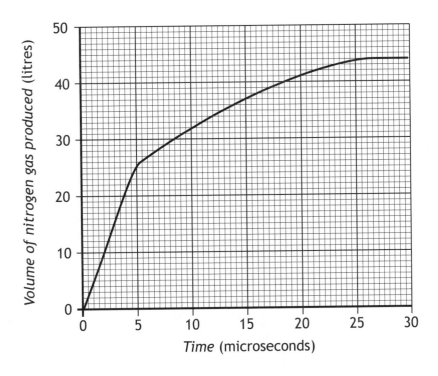

State the total volume, in litres, of nitrogen gas produced.   1

[Turn over

MARKS | DO NOT WRITE IN THIS MARGIN

6. Scientists use an instrument called a mass spectrometer to determine the number of isotopes and the percentage of each isotope in a sample of an element.

   (a) When a sample of boron is passed through a mass spectrometer the following graph is obtained.

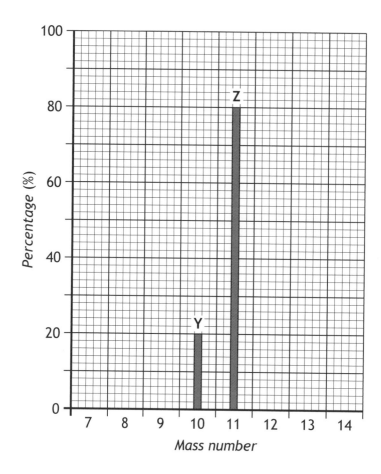

   (i) State the number of isotopes present in this sample of boron.   1

MARKS | DO NOT WRITE IN THIS MARGIN

6. (a) (continued)

   (ii) The relative atomic mass can be calculated using:

$$\frac{(\text{mass of isotope } \mathbf{Y} \times \% \text{ of } \mathbf{Y}) + (\text{mass of isotope } \mathbf{Z} \times \% \text{ of } \mathbf{Z})}{100}$$

   Using the information from the graph, calculate the relative atomic mass of the sample of boron.

   Show your working clearly.

2

   (b) Carbon also has more than one isotope.

   The nuclide notation for an isotope of carbon can be represented as

$$^{12}_{6}C$$

   Write the nuclide notation for the isotope of carbon with 8 neutrons.

1

[Turn over

MARKS | DO NOT WRITE IN THIS MARGIN

7. Strontium chloride, which is an ionic compound, is used in toothpaste to reduce tooth sensitivity.

(a) State the term used to describe the structure of solid strontium chloride.   **1**

(b) A sample of strontium chloride was electrolysed.

(i) State why ionic compounds, like strontium chloride, conduct electricity when molten.   **1**

(ii) During electrolysis, chloride ions lose electrons to form chlorine gas.

Name the type of chemical reaction taking place.   **1**

(iii) Explain why a d.c. supply **must** be used.   **1**

MARKS | DO NOT WRITE IN THIS MARGIN

8. Water is one of the most versatile of all chemicals and features in many chemical reactions and processes.

   Using your knowledge of chemistry, comment on the chemistry of water.   3

[Turn over

MARKS | DO NOT WRITE IN THIS MARGIN

9. Olive oil, which can be used in cooking, is a mixture of unsaturated molecules.

   (a) (i) State what is meant by the term unsaturated.   1

        (ii) Describe the chemical test, including the result, that can be used to show that olive oil is unsaturated.   1

   (b) When frying food, it is recommended that the oil is heated before food is added.

        The table gives information about olive oil used to fry food.

Specific heat capacity of olive oil	$1.97 \, kJ \, kg^{-1} \, {}^{\circ}C^{-1}$
Initial temperature of olive oil	$20 \, {}^{\circ}C$
Mass of olive oil heated	$1500 \, g$

   Calculate the energy, in kJ, required to increase the temperature of the olive oil to $180 \, {}^{\circ}C$.   3

   Show your working clearly.

MARKS | DO NOT WRITE IN THIS MARGIN

10. Ammonia is made industrially by reacting nitrogen with hydrogen.

(a) The equation for this reaction is

$$N_2 \quad + \quad H_2 \quad \rightleftharpoons \quad NH_3$$

(i) Balance the equation above.   **1**

(ii) In the equation the symbol $\rightleftharpoons$ is used.
State what this indicates about the reaction.   **1**

(b) Draw a diagram, showing **all** outer electrons, to represent a molecule of ammonia, NH_3.   **1**

(c) In industry, ammonia can be converted into nitric acid.
Name this industrial process.   **1**

(d) Ammonia reacts with nitric acid to produce a salt.
Name the salt produced in this reaction.   **1**

MARKS | DO NOT WRITE IN THIS MARGIN

11. A student set up the following cell.

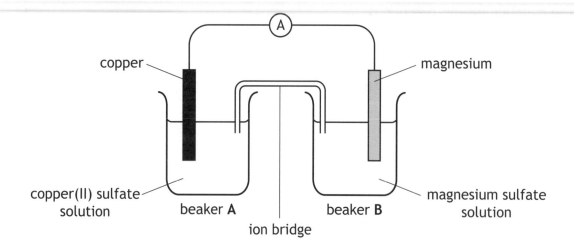

(a) **On the diagram**, draw an arrow to show the path and direction of electron flow.   **1**

You may wish to use the data booklet to help you.

(b) Explain why an ion bridge is used to link the beakers.   **1**

(c) In this reaction, the copper ions are reduced.

Write the ion-electron equation for the reduction of copper(II) ions.   **1**

You may wish to use the data booklet to help you.

MARKS | DO NOT WRITE IN THIS MARGIN

11. **(continued)**

(d) Other magnesium compounds could be used in place of magnesium sulfate when making this type of cell.

Suggest why magnesium phosphate would **not** be suitable.   1

You may wish to use the data booklet to help you.

[Turn over

MARKS | DO NOT WRITE IN THIS MARGIN

12. Thallium-204 decays by emitting beta particles and can be used in industry to measure the thickness of paper.

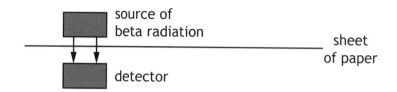

source of beta radiation

sheet of paper

detector

(a) Suggest a reason why a radioisotope which emits alpha particles is not suitable for this purpose.

1

(b) A paper manufacturer found a thallium-204 source had only $\frac{1}{16}$ of its original activity.

The half-life of thallium-204 is 3·7 years.

Calculate the age, in years, of the source.

Show your working clearly.

2

(c) Circle the correct words to complete the sentence.

1

When an atom emits a beta particle,

the atomic number of the atom
$$\left\{ \begin{array}{l} \text{increases} \\ \text{decreases} \\ \text{stays the same} \end{array} \right\}$$ and

the mass number
$$\left\{ \begin{array}{l} \text{increases} \\ \text{decreases} \\ \text{stays the same} \end{array} \right\}.$$

MARKS | DO NOT WRITE IN THIS MARGIN

13. Malic acid is a carboxylic acid found in some fruits.

(a) (i) Name the functional group circled in the diagram above. **1**

(ii) Calculate the mass, in grams, of 1 mole of malic acid. **1**

[Turn over

MARKS | DO NOT WRITE IN THIS MARGIN

13.   (continued)

(b) Carboxylic acids can contain a halogen atom. The pH of $1\,mol\,l^{-1}$ solutions of some of these acids are given in the table.

Carboxylic acid	pH
Br—C—C—OH (H, H, O)	1·45
Cl—C—C—OH (H, H, O)	1·42
F—C—C—OH (H, H, O)	1·33
I—C—C—OH (H, H, O)	1·55

Describe how the acidity of the carboxylic acid is related to the position of the halogen in group 7 of the periodic table.   1

MARKS | DO NOT WRITE IN THIS MARGIN

13. **(continued)**

(c) The Jones oxidation reaction can be used to convert alcohols to carboxylic acids.

The following alcohol can also be converted to a carboxylic acid by the Jones oxidation reaction.

Draw a structural formula for the carboxylic acid produced in this reaction.

1

[Turn over

MARKS | DO NOT WRITE IN THIS MARGIN

14. Chloride ion concentrations greater than $0.25\,g\,l^{-1}$ can cause a noticeable taste in drinking water.

The table gives information about the chloride ion concentration in drinking water from different sources.

Source	Chloride ion concentration $(g\,l^{-1})$
A	0.26
B	0.28
C	0.24

(a) One of the sources provides drinking water that does **not** have a noticeable taste.

Identify this source.

1

(b) A student investigated the concentration of chloride ions in drinking water from another source.

Samples of water were titrated with silver nitrate solution.

An indicator was used to show when the end-point was reached.

$0.02\,mol\,l^{-1}$ silver nitrate solution

$20\,cm^3$ sample of water + indicator

Titration	Volume of silver nitrate added (cm^3)
1	9.6
2	8.0
3	8.5
4	8.1

MARKS | DO NOT WRITE IN THIS MARGIN

14. **(b)** **(continued)**

(i) Name the most appropriate piece of apparatus to measure $20\,cm^3$ samples of water into the flask.

1

(ii) The average volume of silver nitrate that should be used to calculate the chloride ion concentration is $8\cdot05\,cm^3$.

Explain why only the results of titration **2** and titration **4** are used to calculate this average.

1

(iii) Calculate the number of moles of silver nitrate in $8\cdot05\,cm^3$.

Show your working clearly.

1

[Turn over

15. Read the passage and answer the questions that follow.

The bizarre world of high pressure chemistry

What would happen if you put some sodium, normally a soft grey metal, under extremely high pressure?

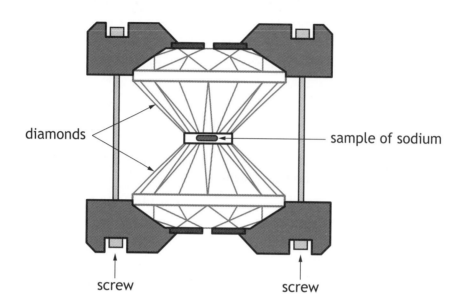

Researchers investigated this using a piece of apparatus called a diamond anvil cell. The diamond anvil cell contains two diamonds and as the screws are tightened, high pressure is created. The pressure between the diamonds can reach 1000 gigapascals, which is a pressure of 10 million atmospheres.

When sodium is squeezed to 190 gigapascals it loses an important property of metals and becomes an insulator. This shows that there is a change in the structure and bonding of sodium.

The diamond anvil cell also allows scientists to create new materials, including superconductors.

Scientists are studying what happens to the materials thought to be deep inside the Earth, where high pressure occurs naturally.

Using this technique to mirror what may happen to materials deep in the Earth, iron(III) oxide is found to decompose, releasing oxygen, and forming the very unusual Fe_5O_7.

Adapted from *The Catalyst*, Volume 27, Number 1, October 2016

MARKS | DO NOT WRITE IN THIS MARGIN

15. (continued)

(a) Name the piece of apparatus used by researchers to create high pressure. **1**

(b) (i) Calculate the pressure, in atmospheres, when sodium is squeezed at 190 gigapascals. **1**

(ii) Suggest what would be observed if this pressurised sodium was placed in the circuit below. **1**

sample of pressurised sodium

(c) Write an equation, using symbols and formulae, to show the decomposition of iron(III) oxide, at high pressure. **1**

There is no need to balance this equation.

MARKS | DO NOT WRITE IN THIS MARGIN

16. The thiols are a family of compounds containing carbon, hydrogen and sulfur.

Name	Full structural formula
methanethiol	
ethanethiol	
propanethiol	

(a) Thiols have the same general formula and similar chemical properties.

(i) State the term used to describe a family of compounds such as the thiols.

1

(ii) Suggest a general formula for this family.

1

(b) Ethanethiol can react with oxygen as shown.

ethanethiol + oxygen → carbon dioxide + water + **Y**

Identify **Y**.

1

MARKS | DO NOT WRITE IN THIS MARGIN

16. **(continued)**

(c) Methanethiol, which smells like rotting cabbage, is added to natural gas to allow gas leaks to be detected.

It is prepared industrially by the reaction of methanol with hydrogen sulfide gas.

$$CH_3OH \ + \ H_2S \longrightarrow CH_3SH \ + \ H_2O$$

Calculate the mass of methanethiol, in grams, produced when 640 grams of methanol reacts completely with hydrogen sulfide.

3

Show your working clearly.

MARKS | DO NOT WRITE IN THIS MARGIN

17. Methacrylic acid is used to make methacrylates which are used in Shellac nail polish.

methacrylic acid

Using your knowledge of chemistry, comment on the chemistry of methacrylic acid.

3

[END OF QUESTION PAPER]

ADDITIONAL SPACE FOR ANSWERS AND ROUGH WORKING

Additional graph paper for Question 1 (b) (ii)

ADDITIONAL SPACE FOR ANSWERS AND ROUGH WORKING

NATIONAL 5

2019

National Qualifications 2019

X813/75/02

Chemistry
Section 1—Questions

FRIDAY, 10 MAY

1:00 PM — 3:30 PM

Instructions for the completion of Section 1 are given on *Page two* of your question and answer booklet X813/75/01.

Record your answers on the answer grid on *Page three* of your question and answer booklet.

You may refer to the Chemistry Data Booklet for National 5.

Before leaving the examination room you must give your question and answer booklet to the Invigilator; if you do not, you may lose all the marks for this paper.

SECTION 1 — 25 marks
Attempt ALL questions

Questions 1 and 2 refer to an experiment to investigate the rate of a reaction.

The volume of gas collected in 2 minutes was $5\,cm^3$.

1. What was the average rate of reaction over this time?

   A 0·2

   B 0·4

   C 2·5

   D 5·0

2. The unit for the average rate of this reaction is

   A cm^3/min^{-1}

   B $cm^3\,min^{-1}$

   C min/cm^3

   D $min\,cm^{-3}$

3. Tennessine is a newly discovered element with a predicted electron arrangement of 2,8,18,32,32,18,7.

   In which group of the periodic table should Tennessine be placed?

   A 1

   B 2

   C 7

   D 8

4. Which of the following is a positively charged ion?

	Protons	Neutrons	Electrons
A	9	10	10
B	10	9	10
C	11	12	11
D	12	13	10

5. To turn a gas into a liquid it must be cooled below a temperature known as its critical temperature.

Gas	Formula	Relative formula mass	Critical temperature (°C)
hydrogen	H_2	2	−240
helium	He	4	−268
ammonia	NH_3	17	133
oxygen	O_2	32	−119
carbon dioxide	CO_2	44	31

Identify the true statement based on the information in this table.

A   Carbon dioxide can be a liquid at 40 °C

B   Compounds have higher critical temperatures than elements

C   Critical temperature increases as relative formula mass increases

D   Diatomic elements have lower critical temperatures than Noble gases

6. A molecule of phosphorus trifluoride is shown.

Which term can be used to describe the shape of a phosphorus trifluoride molecule?

A   Linear

B   Angular

C   Tetrahedral

D   Trigonal pyramidal

[Turn over

7. In which of the following compounds do the ions have the same electron arrangement? You may wish to use the data booklet to help you.

   A Na_2O

   B   LiF

   C   KBr

   D $MgCl_2$

8. Several conductivity experiments were carried out using the apparatus below.

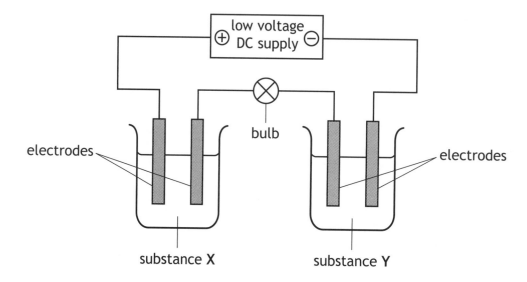

   Identify the experiment in which the bulb would light.

	Substance X	Substance Y
A	solid copper sulfate	liquid mercury
B	copper chloride solution	molten sodium chloride
C	solid potassium nitrate	nickel bromide solution
D	sodium chloride solution	liquid hexane

9. Limewater can be made by dissolving calcium hydroxide in water.

   Which of the following terms correctly describes calcium hydroxide?

   A   Solute

   B   Solvent

   C   Solution

   D   Insoluble

10. Ammonium nitrate, NH_4NO_3, has a gram formula mass of 80.

    The percentage by mass of nitrogen in ammonium nitrate is equal to

    A $\frac{14}{80} \times 100$

    B $\frac{28}{80} \times 100$

    C $\frac{28}{100} \times 80$

    D $\frac{80}{28} \times 100$.

11. As an alkaline solution is diluted with water

    A   the pH increases

    B   the pH stays the same

    C   the concentration of hydroxide ions increases

    D   the concentration of hydroxide ions decreases.

12. Which of the following compounds is a base?

    A   Sodium oxide

    B   Calcium chloride

    C   Potassium nitrate

    D   Ammonium sulfate

[Turn over

13. Which of the following compounds does **not** have an isomer?

    A   Cyclopropane

    B   But-1-ene

    C   Pentane

    D   Ethene

14. The systematic name for $CH_3CH_2C(CH_3)CHCH_3$ is

    A   3-methylpentane

    B   2-methylpentane

    C   3-methylpent-2-ene

    D   2-methylpent-3-ene.

15. When pent-1-ene undergoes an addition reaction with water, two products are formed.

    Which of the following alkenes will also produce two products when it undergoes an addition reaction with water?

    A   Oct-2-ene

    B   Hex-3-ene

    C   But-2-ene

    D   Ethene

16. In the Clemmensen reaction, ketones can be converted to alkanes as shown.

ketone   alkane

Identify the alkane produced if the following ketone was used in this reaction?

ketone

A

B

C

D

[Turn over

17. Which line in the table correctly describes the trends going from hexanoic acid to butanoic acid?

	Formula mass	Solubility in water
A	increasing	decreasing
B	decreasing	increasing
C	decreasing	decreasing
D	increasing	increasing

18. Four cells were made by joining silver to different metals.

The cells produced the following voltages 2·7 V, 1·1 V, 0·9 V and 0·5 V.

The metals used were copper, zinc, iron and magnesium.

Which voltage was produced in the cell containing silver and copper?

You may wish to use the data booklet to help you.

A   2·7 V

B   1·1 V

C   0·9 V

D   0·5 V

19. Information about the reactions of three different metals, **X**, **Y** and **Z** is given in the table.

Metal	Reaction with dilute acid	Reaction with water
X	reacts	no reaction
Y	no reaction	no reaction
Z	reacts	reacts

Which of the following shows the metals in order of **increasing** reactivity?

A   Y, Z, X

B   Z, X, Y

C   Y, X, Z

D   X, Y, Z

20. A co-polymer is formed when two different monomers polymerise.

Part of the structure of a co-polymer, showing three monomer units, is given below.

One of the monomers used is propene.

Identify the other monomer.

A   Pent-2-ene

B   Pent-1-ene

C   But-2-ene

D   But-1-ene

[Turn over

21. Nitrogen dioxide is a brown coloured gas that is soluble in water and more dense than air.

Which of the following diagrams shows the most appropriate method for collecting and measuring the volume of nitrogen dioxide?

A

measuring cylinder

water

B

measuring cylinder

tub

water

C

measuring cylinder

D

measuring cylinder

22. A solution of a metal chloride burns with a green flame.

    Which of the following metal ions could be present in the metal chloride?

    You may wish to use the data booklet to help you.

    A Ba^{2+}

    B Ca^{2+}

    C K^+

    D Na^+

23. Identify the gas that turns limewater cloudy.

    A   Oxygen

    B   Nitrogen

    C   Hydrogen

    D   Carbon dioxide

    Questions **24** and **25** refer to the equation shown.

$$AgNO_3(aq) \;+\; NaBr(aq) \;\rightarrow\; NaNO_3(aq) \;+\; AgBr(s)$$

24. The reaction shown by the equation is an example of

    A   addition

    B   combustion

    C   precipitation

    D   neutralisation.

25. Which of the following ions are spectator ions in the reaction?

    A Ag^+ and NO_3^-

    B Na^+ and NO_3^-

    C Ag^+ and Br^-

    D Na^+ and Br^-

[END OF SECTION 1. NOW ATTEMPT THE QUESTIONS IN SECTION 2 OF YOUR QUESTION AND ANSWER BOOKLET]

[BLANK PAGE]

DO NOT WRITE ON THIS PAGE

N5

National Qualifications 2019

Mark

X813/75/01

**Chemistry
Section 1—Answer Grid
and Section 2**

FRIDAY, 10 MAY

1:00 PM — 3:30 PM

Fill in these boxes and read what is printed below.

Full name of centre

Town

Forename(s)

Surname

Number of seat

Date of birth

Day   Month   Year   Scottish candidate number

Total marks — 100

SECTION 1 — 25 marks

Attempt ALL questions.

Instructions for the completion of Section 1 are given on *Page two*.

SECTION 2 — 75 marks

Attempt ALL questions.

You may refer to the Chemistry Data Booklet for National 5.

Write your answers clearly in the spaces provided in this booklet. Additional space for answers and rough work is provided at the end of this booklet. If you use this space you must clearly identify the question number you are attempting. Any rough work must be written in this booklet. You should score through your rough work when you have written your final copy.

Use **blue** or **black** ink.

Before leaving the examination room you must give this booklet to the Invigilator; if you do not, you may lose all the marks for this paper.

SECTION 1 — 25 marks

The questions for Section 1 are contained in the question paper X813/75/02.

Read these and record your answers on the answer grid on *Page three* opposite.

Use **blue** or **black** ink. Do NOT use gel pens or pencil.

1. The answer to each question is **either** A, B, C or D. Decide what your answer is, then fill in the appropriate bubble (see sample question below).

2. There is **only one correct** answer to each question.

3. Any rough working should be done on the additional space for answers and rough work at the end of this booklet.

Sample question

To show that the ink in a ball-pen consists of a mixture of dyes, the method of separation would be

    A   fractional distillation

    B   chromatography

    C   fractional crystallisation

    D   filtration.

The correct answer is **B** — chromatography. The answer **B** bubble has been clearly filled in (see below).

Changing an answer

If you decide to change your answer, cancel your first answer by putting a cross through it (see below) and fill in the answer you want. The answer below has been changed to **D**.

If you then decide to change back to an answer you have already scored out, put a tick (✓) to the **right** of the answer you want, as shown below:

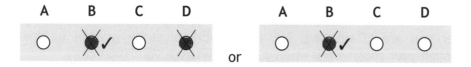

or

SECTION 1 — Answer Grid

	A	B	C	D
1	○	○	○	○
2	○	○	○	○
3	○	○	○	○
4	○	○	○	○
5	○	○	○	○
6	○	○	○	○
7	○	○	○	○
8	○	○	○	○
9	○	○	○	○
10	○	○	○	○
11	○	○	○	○
12	○	○	○	○
13	○	○	○	○
14	○	○	○	○
15	○	○	○	○
16	○	○	○	○
17	○	○	○	○
18	○	○	○	○
19	○	○	○	○
20	○	○	○	○
21	○	○	○	○
22	○	○	○	○
23	○	○	○	○
24	○	○	○	○
25	○	○	○	○

[BLANK PAGE]

DO NOT WRITE ON THIS PAGE

[Turn over for next question

DO NOT WRITE ON THIS PAGE

MARKS | DO NOT WRITE IN THIS MARGIN

SECTION 2 — 75 marks

Attempt ALL questions

1. There are many different types of glass.

    Glass is made from the chemical silica, SiO_2, which is obtained from sand.

    (a) Silica has a melting point of 1713 °C.

    State the term used to describe the structure of silica.   **1**

    (b) Borosilicate glass is a type of glass that also contains the element boron.

    A sample of boron contains two different types of atom.

    $$^{10}_{5}B \qquad ^{11}_{5}B$$

    (i) State the term used to describe these different types of boron atom.   **1**

    (ii) Explain why the mass number of each type of boron atom is different.   **1**

    (iii) The relative atomic mass of boron is 10·8.

    State the mass number of the most common type of atom in the sample.   **1**

MARKS | DO NOT WRITE IN THIS MARGIN

1. (continued)

(c) Glass that contains a minimum of 24% lead oxide is known as crystal glass.

Calculate the mass, in grams, of lead oxide in a sample of crystal glass weighing 500 g.

1

[Turn over

MARKS | DO NOT WRITE IN THIS MARGIN

2. Read the passage below and answer the questions that follow.

Antifreeze

Antifreeze lowers the freezing point of water. When diluted, antifreeze is used in car engines to prevent water-based liquids from freezing.

Different brands of antifreeze can contain either ethane-1,2-diol or propane-1,2-diol.

ethane-1,2-diol   propane-1,2-diol

Ethane-1,2-diol is toxic if swallowed. In the liver, an enzyme converts ethane-1,2-diol into oxalic acid.

oxalic acid

The oxalic acid then reacts with calcium in the body to form calcium oxalate. Calcium oxalate is the main component of kidney stones, which can cause extreme pain.

Propane-1,2-diol is not regarded as toxic because the body breaks down the molecule to harmless lactic acid, which is also produced naturally in the body during exercise.

lactic acid

Adapted from *Education in Chemistry, May 2008, Volume 45 Number 3*

MARKS | DO NOT WRITE IN THIS MARGIN

2. **(continued)**

(a) Name the functional group found in both ethane-1,2-diol and propane-1,2-diol.

1

(b) Name the type of substance used to convert ethane-1,2-diol into oxalic acid.

1

(c) Name the salt mentioned in the passage.

1

(d) Calculate the mass, in grams, of 1 mole of the harmless product formed, in the body, from propane-1,2-diol.

1

[Turn over

MARKS | DO NOT WRITE IN THIS MARGIN

3. Nitrogen and hydrogen react together to form ammonia.

$$N_2(g) \ + \ 3H_2(g) \ \rightleftharpoons \ 2NH_3(g)$$

(a) Draw a diagram, showing **all** outer electrons, to represent a molecule of nitrogen gas, N_2.

1

(b) The following method can be used to prepare small quantities of ammonia in the laboratory.

Suggest what colour the damp pH paper would be after the mixture is heated.

1

MARKS | DO NOT WRITE IN THIS MARGIN

3. (continued)

(c) In industry, ammonia can be produced by the Haber process.

The table shows the yield of ammonia produced at different temperatures by this process.

Temperature (°C)	100	200	400	500	600	700
Percentage yield of ammonia (%)	97	87	46	28	17	10

(i) Describe the relationship between temperature and percentage yield of ammonia.

1

(ii) Draw a graph of the percentage yield of ammonia against temperature.

4

(Additional graph paper, if required, can be found on *Page thirty-three*.)

MARKS | DO NOT WRITE IN THIS MARGIN

3. **(continued)**

(d) Scientists are developing an alternative industrial process to produce ammonia, which is more efficient than the Haber process.

This involves the electrolysis of molten lithium hydroxide to produce lithium, water and oxygen. Lithium is then reacted with nitrogen gas, which is obtained from air, to produce lithium nitride. Ammonia and lithium hydroxide are produced when lithium nitride reacts with water.

(i) Complete the flow diagram using the information above. 1

(An additional diagram, if required, can be found on *Page thirty-four*.)

(ii) **On the flow diagram**, draw an arrow to show how the process can be made more economical. 1

MARKS | DO NOT WRITE IN THIS MARGIN

4. Radioisotopes emit radiation to become more stable.

(a) State where the radioactive decay occurs in an atom. **1**

(b) Iodine-131 is a radioisotope with a half-life of 8 days and can be used in the treatment of thyroid cancer.

   (i) State what is meant by the term half-life. **1**

   (ii) Calculate the percentage of iodine-131 that would have **decayed** after 24 days. **3**

   (iii) Different concentrations of iodine-131 are used to treat different types of cancer.

   Circle the correct words to complete the sentence. **1**

   When an iodine-131 solution is diluted,

   the half-life { gets longer / stays the same / gets shorter }.

MARKS | DO NOT WRITE IN THIS MARGIN

5. The alkenes are a family of unsaturated hydrocarbons.

(a) Describe the chemical test, including the result, to show that a hydrocarbon is unsaturated.

1

(b) Propene is an alkene that can take part in a range of addition reactions.

(i) Name the type of addition reaction taking place in reaction **X**.

1

(ii) Name the chemical that reacts with propene to form compound **Y**.

1

(iii) Name the polymer formed in reaction **Z**.

1

MARKS | DO NOT WRITE IN THIS MARGIN

5. (continued)

(c) The cycloalkenes are another family of unsaturated hydrocarbons.

(i) Cyclohexene can be made by reacting ethene with butadiene in a reaction called the Diels-Alder reaction as shown.

$$C_2H_4 \quad + \quad C_4H_6 \quad \rightarrow \quad C_6H_{10}$$

ethene   butadiene   cyclohexene

Calculate the mass, in grams, of ethene required to make 410 g of cyclohexene.

Show your working clearly.

3

(ii) The table gives information about cyclopentene and cyclohexene.

Cycloalkene	Boiling point (°C)
cyclopentene	45
cyclohexene	83

Explain why cyclopentene has a lower boiling point than cyclohexene.

2

MARKS | DO NOT WRITE IN THIS MARGIN

6. An oxide is a compound that contains at least one oxygen atom and **only one** other element in its chemical formula.

   Using your knowledge of chemistry, comment on the chemistry of oxides.   3

[Turn over for next question

DO NOT WRITE ON THIS PAGE

MARKS | DO NOT WRITE IN THIS MARGIN

7. Paraffin wax is a mixture of hydrocarbon molecules that belong to the same homologous series.

(a) State what is meant by the term homologous series.

1

(b) An example of one hydrocarbon contained in paraffin wax is $C_{25}H_{52}$.

(i) Name the homologous series to which this hydrocarbon belongs.

1

(ii) Write the molecular formula for the molecule, containing 72 hydrogen atoms, that belongs to the same homologous series.

1

MARKS | DO NOT WRITE IN THIS MARGIN

7. (continued)

(c) The table contains information about some hydrocarbon molecules.

Number of carbon atoms	Boiling point (°C)
20	343
21	356
22	369
23	381

Predict the boiling point, in °C, of the hydrocarbon with 24 carbon atoms.   1

[Turn over

MARKS | DO NOT WRITE IN THIS MARGIN

8. Read the passage below and answer the questions that follow.

Beryllium

Beryllium is a rare element in the universe. Unlike most elements it was not formed during the Big Bang or by stars. In fact, beryllium is only formed in supernova explosions.

Beryllium is found in the mineral Beryl, which has the chemical name beryllium aluminium silicate. Beryl makes up a range of glittering gemstones such as emerald and aquamarine.

In 1828 the metal beryllium was extracted from beryllium chloride ($BeCl_2$) by reacting this compound with potassium. Potassium chloride was also produced in this reaction.

In 1932 James Chadwick discovered when a sample of beryllium was bombarded with X-rays from radium, it emitted a new kind of sub-atomic particle that had mass but no charge. He called this new particle a neutron and was awarded the Nobel Prize for his work in 1935.

Adapted from Education in Chemistry, November 2015, Volume 52, Issue 6

(a) State where beryllium is formed.   1

(b) Name the elements found in the mineral Beryl.   1

MARKS | DO NOT WRITE IN THIS MARGIN

8. (continued)

(c) Write an equation, using symbols and formulae, to show the reaction taking place when beryllium is extracted from beryllium chloride.

There is no need to balance the equation.   **1**

(d) During the extraction of beryllium, the beryllium ions are changed to beryllium atoms.

Name this type of chemical reaction.   **1**

(e) Write the nuclide notation for the sub-atomic particle discovered by James Chadwick in 1932.   **1**

[Turn over

MARKS | DO NOT WRITE IN THIS MARGIN

9. Alcohols can take part in different types of chemical reaction.

(a) When alcohols are burned, heat energy is released.

State the term used to describe all chemical reactions that release heat energy.

1

(b) A student carried out the following experiment.

thermometer

glass beaker

ethanol

(i) When 0·8 g of ethanol was burned, 8·36 kJ of energy was absorbed by the water.

If the temperature of the water increased by 40 °C, calculate the mass, in kg, of water used by the student in this experiment.

3

You may wish to use the data booklet to help you.

Show your working clearly.

MARKS | DO NOT WRITE IN THIS MARGIN

9. (b) (continued)

(ii) The experiment was repeated, replacing the glass beaker with a copper can and using a heat shield.

Explain why these changes resulted in more heat energy being absorbed by the water.

2

Improvement	Explanation
Use of a copper can	
Use of a heat shield	

[Turn over

MARKS | DO NOT WRITE IN THIS MARGIN

9. (continued)

(c) Alcohols can react with hot copper(II) oxide.

Depending on the structure of the alcohol used, the product will be either an aldehyde or a ketone.

Structural formula of alcohol	Type of product	Structural formula of product
	aldehyde	
	aldehyde	
	ketone	
	ketone	
	ketone	

(i) Write a general statement linking the position of the functional group in an alcohol to the **type** of product formed.

1

MARKS | DO NOT WRITE IN THIS MARGIN

9. (c) (continued)

(ii) The following alcohol reacts with hot copper(II) oxide to produce an aldehyde.

$$H-\overset{\displaystyle H}{\underset{\displaystyle H}{C}}-\overset{\displaystyle H}{\underset{\displaystyle H}{C}}-\overset{\displaystyle H}{\underset{\displaystyle H}{C}}-\overset{\displaystyle H}{\underset{\displaystyle H}{C}}-\overset{\displaystyle H}{\underset{\displaystyle H}{C}}-OH$$

Draw the full structural formula for the aldehyde produced when this alcohol reacts with hot copper(II) oxide.

1

[Turn over

MARKS | DO NOT WRITE IN THIS MARGIN

10. A student set up an electrochemical cell using solutions of iron(III) chloride and potassium iodide.

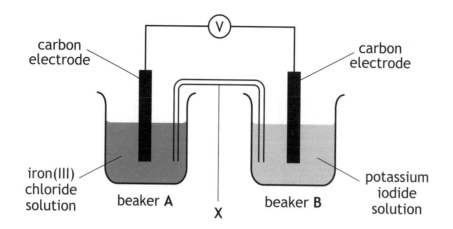

The reactions taking place are

beaker **A** $Fe^{3+}(aq) + e^- \rightarrow Fe^{2+}(aq)$

beaker **B** $2I^-(aq) \rightarrow I_2(\ell) + 2e^-$

(a) Name the piece of apparatus labelled **X**.   1

(b) (i) **On the diagram**, draw an arrow to show the path and direction of electron flow.   1

You may wish to use the data booklet to help you.

(ii) Name the type of chemical reaction taking place in beaker **B**.   1

MARKS | DO NOT WRITE IN THIS MARGIN

10. (b) (continued)

(iii) Write the redox equation for the overall reaction.   1

(c) Carbon in the form of graphite is a suitable material for use as an electrode as it does not react with the solutions.

Suggest another reason why it is a suitable material.   1

[Turn over

MARKS | DO NOT WRITE IN THIS MARGIN

11. A student was asked to prepare the soluble compound, calcium propanoate.

A section of the procedure used by the student is shown.

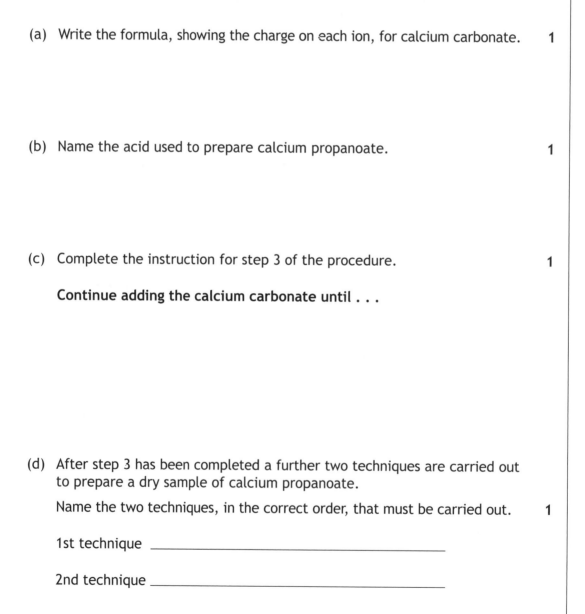

Preparation of calcium propanoate

Procedure

1. Using a measuring cylinder add 20 cm³ of dilute acid to a beaker.
2. Add a spatulaful of calcium carbonate to the acid and stir the reaction mixture with a glass rod.
3. Continue adding the calcium carbonate until . . .

(a) Write the formula, showing the charge on each ion, for calcium carbonate.   1

(b) Name the acid used to prepare calcium propanoate.   1

(c) Complete the instruction for step 3 of the procedure.   1

Continue adding the calcium carbonate until . . .

(d) After step 3 has been completed a further two techniques are carried out to prepare a dry sample of calcium propanoate.

Name the two techniques, in the correct order, that must be carried out.   1

1st technique _____

2nd technique _____

MARKS | DO NOT WRITE IN THIS MARGIN

12. A student carried out a titration experiment to calculate the concentration of a solution of hydrochloric acid.

(a) Before the titration was carried out the student prepared a 200 cm^3 solution of sodium carbonate.

This solution had an accurate concentration of 1·0 mol l^{-1}.

    (i) State the term given to a solution of accurately known concentration.　　**1**

    (ii) Calculate the mass, in grams, of sodium carbonate, Na$_2$CO$_3$, required to prepare 200 cm^3 of 1·0 mol l^{-1} solution.　　**3**

Show your working clearly.

[Turn over

MARKS | DO NOT WRITE IN THIS MARGIN

12.   (continued)

(b) The student performed the titration as shown.

1·0 mol l^{-1} sodium carbonate solution

20·0 cm^3 hydrochloric acid solution and indicator

(i)   Suggest one improvement to the student's experimental technique.   1

(ii) State why an indicator is used.   1

MARKS | DO NOT WRITE IN THIS MARGIN

12.　(continued)

(iii)　The average volume of sodium carbonate used was $15 \cdot 0 \, cm^3$.

To calculate the average volume of sodium carbonate used, the student only used titre volumes within $0 \cdot 2 \, cm^3$ of each other.

State the term used to describe these titre volumes.　**1**

(iv)　The equation for the reaction is

$$Na_2CO_3(aq) \ + \ 2HCl(aq) \ \rightarrow \ 2NaCl(aq) \ + \ H_2O(\ell) \ + \ CO_2(g)$$

Calculate the concentration, in $mol \, l^{-1}$, of the hydrochloric acid solution.　**3**

Show your working clearly.

[Turn over for next question

MARKS | DO NOT WRITE IN THIS MARGIN

13. The force of attraction between oppositely charged particles is important in chemistry.

Using your knowledge of chemistry, explain why this force of attraction is important.

3

[END OF QUESTION PAPER]

MARKS | DO NOT WRITE IN THIS MARGIN

ADDITIONAL SPACE FOR ANSWERS AND ROUGH WORK

Additional graph for use with question 3 (c) (ii)

MARKS | DO NOT WRITE IN THIS MARGIN

ADDITIONAL SPACE FOR ANSWERS AND ROUGH WORK

Additional diagram for question 3 (d) (i)

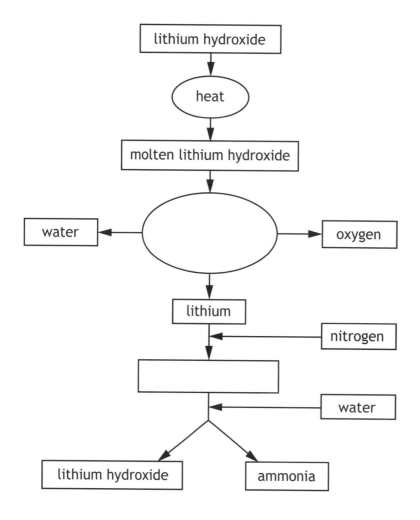

MARKS | DO NOT WRITE IN THIS MARGIN

ADDITIONAL SPACE FOR ANSWERS AND ROUGH WORKING

MARKS | DO NOT WRITE IN THIS MARGIN

ADDITIONAL SPACE FOR ANSWERS AND ROUGH WORKING

NATIONAL 5

Answers

SQA NATIONAL 5 CHEMISTRY 2019

NATIONAL 5 CHEMISTRY 2017 SPECIMEN QUESTION PAPER

Section 1

Question	Answer	Max mark
1.	B	1
2.	C	1
3.	A	1
4.	D	1
5.	A	1
6.	D	1
7.	A	1
8.	C	1
9.	C	1
10.	A	1
11.	D	1
12.	A	1
13.	C	1
14.	B	1
15.	A	1
16.	B	1
17.	C	1
18.	D	1
19.	D	1
20.	A	1
21.	B	1
22.	A	1
23.	B	1
24.	C	1
25.	C	1

Section 2

1. (a) 86—90 (seconds)

    Units are not required, but 0 marks should be awarded for the correct answer if incorrect unit is given

   (b) $1 \cdot 5$ cm^3 s^{-1} (3)

    Partial marking:
    $1 \cdot 5$ with no unit/incorrect unit (2)
    $\dfrac{30-0}{20-0}$ or $\dfrac{30}{20}$ or $\dfrac{0-30}{0-20}$ (1)
    Correct unit cm^3 s^{-1} (1)

   (c) Less reactants
    or
    Concentration of reactants decreases
    or
    Reactants are used up
    or
    Less chance of particles colliding
    or
    Equivalent answer

2. (a) Atoms with same atomic number/number of protons/positive particles but different mass number/number of neutrons

   (b) Protons = 35
    Neutrons = 44
    Both required for 1 mark

   (c) Equal amounts/proportions/abundance
    or
    Same number of each
    or
    50:50
    or
    Equivalent answers

   (d) This is an open-ended question.
    1 mark: The candidate has demonstrated a limited understanding of the chemistry involved. The candidate has made a/some statement(s) which is/are relevant to the situation, showing that at least a little of the chemistry within the problem is understood.
    2 marks: The candidate has demonstrated a reasonable understanding of the chemistry involved. The candidate has made a/some statement(s) which is/are relevant to the situation, showing that the problem is understood.
    3 marks: The candidate has demonstrated a good understanding of the chemistry involved. The candidate shows a good comprehension of the chemistry of the situation and has provided a logically correct answer to the question posed. This type of response might include a statement of the principles involved, a relationship or an equation, and the application of these to respond to the problem. This does not mean the answer has to be what might be termed an "excellent" answer or a "complete" one.

3. (a) $Al^{3+}(OH^-)_3$

   (b) (i) $11 \cdot 2$ (pence).

    Unit is not required, but 0 marks can be awarded for the correct answer if incorrect unit is given. Incorrect units would only be penalised once in any paper

    (ii) Named active ingredient with an appropriate reason.

    e.g. magnesium hydroxide
    — cheapest/doesn't fizz

    Aluminium hydroxide — need to take least amount

4. (a) (a higher)

   (b) (i) For appropriate format: scatter graph — i.e. a graph in which points are plotted with their x and y values representing temperature and solubility (1)

    The axis/axes of the graph has/have suitable scale(s). For the graph paper provided within the question paper, the selection of suitable scales will result in a graph that occupies at least half of the width and half of the height of the graph paper (1)

    The axes of the graph have suitable labels and units (1)

All data points plotted accurately with a line of best fit drawn (1)

(ii) $10\cdot2 - 10\cdot3$ (g/100 cm^3)

or

A value correctly read from candidate's graph (allow ½ box tolerance)

Units are not required, but 0 marks can be awarded for correct answer if incorrect unit is given

5. (a) (i) $Li_2CO_3 + 2HCl \rightarrow 2LiCl + CO_2 + H_2O$
(or correct multiples)

(ii) LiCl
or
Lithium chloride

(b) (i) $0\cdot44$ (g) (3)

Units are not required, but a maximum of 2 marks can be awarded for the correct answer if incorrect unit is given

Partial marking:
Both *GFMs* 100 and 44 (1)

Correct application of the relationship between moles and mass (1)

This could be shown:

- by working containing the two expressions

$$\frac{1}{candidate's\ GFM\ for\ CaCO_3}$$

and

no. moles CO_2 × candidate's GFM CO_2

or

- by working showing correct proportionality

$$1 \leftrightarrow \frac{candidate\ GFM\ CO_2}{candidate\ GFM\ CaCO_3}$$

Where the candidate has been awarded the mark for the correct application of the relationship between moles and mass, a further mark can be awarded for correct follow through to a final answer (1)

(ii) Method B (1)

Gas is lost in method A before starting mass taken
or
Gas is lost before all acid is added
or
No total mass of all reactants at the start of experiment
or
Equivalent response (1)

6. (a) Flame test (or correct description)
and
lilac/purple
Both required for 1 mark

(b) Greater than 7
or
Any numerical value greater than seven

(c) Ethene
Accept correct formula

(d) Benzoic acid

(e) Aluminium, silicon and oxygen
Accept correct formulae

7. (a) Group/family/chemicals/compounds with same general formula and same/similar chemical properties

Both parts required for 1 mark

(b) (i) Diagram showing carbon with four hydrogen atoms: each of the four overlap areas must have two electrons in or on overlap area (cross, dot, petal diagram)

e.g.

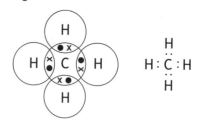

(ii) Weak bond/attraction (1)
between molecules (1)

8. (a) Hydroxyl

(b) Carbon dioxide
and
water
Both required for 1 mark
Accept correct formulae

(c) (i) Addition
or
Hydration

(ii)

```
       H   H   H   H
       |   |   |   |
   H — C — C — C — C — H
       |   |   |   |
       H   |   H   O — H
           |
       H — C — H
           |
           H
```

or

```
       H   H   H           H
       |   |   |           |
   H — C — C — C —————————— C — H
       |   |   |           |
       H   |   O — H        H
           |
       H — C — H
           |
           H
```

Accept full or shortened structural formula

(d) (i)

```
       H         O
       |        //
   H — C — C
       |        \
       H         O — H
```

Accept full or shortened structural formula

(ii) Carboxylic acid
Accept alkanoic acid

9. (a) Exothermic

   (b) (i) 33·44 (kJ) (3)

   Partial marking:
   Using $cm\Delta T$ with $c = 4·18$ (1)

   To be awarded this concept mark, candidates do not specifically need to write $cm\Delta T$. The concept mark is awarded for using this relationship with three values, one of which must be 4·18

   For values 0·2 (kg) and 40 (°C) (1)

   A further mark can be awarded for arithmetical follow through to the candidate's answer only if the mark for the $cm\Delta T$ concept has been awarded (1)

   Units are not required, but a maximum of 2 marks can be awarded for the correct answer if incorrect unit is given

   (ii) Draught insulation
   or
   Use metal beaker
   or
   Repeat to get average
   or
   Any reasonable answer

   (c) (i) As the number of carbons increases the energy released increases
   or
   As the number of carbons decreases the energy released decreases
   or
   The energy increases as the number of carbons increases
   or
   The energy decreases as the number of carbons decreases

   (ii) 3520 to 3550 (kJ)

   Units are not required, 0 marks can be awarded for the correct answer if incorrect unit is given

10. (a)

```
        H
        |
    H — C — H
        |
    H   |   H   H
    |   |   |   |
H — C = C — C = C — H
```

   or

```
    H   CH₃  H   H
    |   |    |   |
H — C = C —  C = C — H
```

   In the formula above, the bond to the methyl group must be correctly aligned with the C atom of the group

   (b) Diagram showing delivery tube passing into a test tube which is placed in a water/ice bath

   Delivery tube must extend close enough to the neck of the test tube to ensure the vapour can enter the test tube

   (c) $C_{10}H_{16}Br_4$

11. (a) Reduction

   (b) (i) d.c.

   (ii) Negative — (brown solid formed)
   Positive — (bubbles of gas)
   Both required for one mark

12. (a) 46·67/46·7/47 (3)

   Partial marking:
   GFM = 60 (1)

   $$\frac{28}{\text{candidate's GFM}} \times 100$$ (1)

   Calculation of final answer using the relationship

   $$\% \text{ by mass} = \frac{m}{GFM} \times 100$$ (1)

   (b) (i) Haber (-Bosch)

   (ii) Speeds up reaction
   or
   Less energy/temperature/heat required

   (c) Platinum
   Accept platinum and rhodium (alloy)

13. (a) (i) Burette

   (ii) 16 or 16·0 (cm^3)
   Units are not required; 0 marks can be awarded for the correct answer if incorrect unit is given

   (b) Titration

14. (a) 0·5 (g) (2)
   Units are not required but a maximum of 1 mark can be awarded for the correct answer if incorrect unit is given

   Partial marking:
   1 mark can be awarded for either:
   • 2 half lives
     or
     Mass correctly calculated for an incorrect number of half-lives shown.

   (b) Short half-life
   or
   Would not last long in the body
   or
   Gamma would go right through body
   or
   Equivalent response

   (c) Beta/β/$_{-1}^{0}e$/$_{-1}^{0}\beta$

   The charge on the beta particle does not need to be shown

   Do not accept electron without atomic and mass numbers, i.e. e or e-

15. (a) 0·01 (mol) (2)
   Units are not required but a maximum of 1 mark can be awarded for the correct answer if incorrect unit is given

   Partial marking:
   1 mark can be awarded for either
   • 143·5 g
   or
   • Correctly calculated answer for $\dfrac{1·435}{\text{incorrect GFM}}$

(b) 0·5 (mol l^{-1}) (2)

Units are not required but a maximum of 1 mark can be awarded for the correct answer if incorrect unit is given

Partial marking:

1 mark can be awarded for either

- $\dfrac{0·01}{0·02}$

or

- Correctly calculated answer for
  $\dfrac{0·01}{20}$

If correct relationship is used but volume not converted to litres e.g. 0·01/20 maximum 1 mark

16. This is an open-ended question.

1 mark: The candidate has demonstrated a limited understanding of the chemistry involved. The candidate has made a/some statement(s) which is/are relevant to the situation, showing that at least a little of the chemistry within the problem is understood.

2 marks: The candidate has demonstrated a reasonable understanding of the chemistry involved. The candidate has made a/some statement(s) which is/are relevant to the situation, showing that the problem is understood.

3 marks: The candidate has demonstrated a good understanding of the chemistry involved. The candidate shows a good comprehension of the chemistry of the situation and has provided a logically correct answer to the question posed. This type of response might include a statement of the principles involved, a relationship or an equation, and the application of these to respond to the problem. This does not mean the answer has to be what might be termed an "excellent" answer or a "complete" one.

NATIONAL 5 CHEMISTRY 2018

Section 1

Question	Answer	Mark
1.	A	1
2.	B	1
3.	B	1
4.	D	1
5.	A	1
6.	D	1
7.	C	1
8.	B	1
9.	C	1
10.	D	1
11.	C	1
12.	C	1
13.	A	1
14.	B	1
15.	B	1
16.	C	1
17.	A	1
18.	B	1
19.	A	1
20.	C	1
21.	C	1
22.	D	1
23.	D	1
24.	B	1
25.	D	1

Section 2

1. (a) Carbon dioxide

   (b) (i) 0·5 or $\frac{1}{2}$ cm^3 s^{-1} (3)

   Partial marking:

   0·5 with no unit/incorrect unit (2)

   OR

   1 mark awarded for concept of change in volume/change in time.

   $\dfrac{77 - 62}{50 - 20}$ or $\dfrac{15}{30}$ or $\dfrac{62 - 77}{20 - 50}$ (1)

   Correct unit cm^3 s^{-1} (1)

   This mark is independent of the calculated value.

   The mark for a final answer can only be awarded if the concept of **change in** volume/**change in** time is correct ie incorrect values **from the table** used (subtractions must be shown and volumes chosen **must** correspond to chosen times).

   (ii) One mark is awarded for a graph which shows points plotted rather than bars. (1)

   The axis/axes of the graph has/have suitable scale(s). For the graph paper provided within the question paper, the selection of suitable scales will result in a graph (plotted points) that occupies at least half of the width and half of the height of the graph paper. (1)

   The axes of the graph have suitable labels and units. (1)

   All data points plotted accurately (within a half box tolerance) with either a line of best fit drawn or plots joined.

   This mark can only be accessed if linear scales for both axes have been provided. (1)

   (iii) Answer must be correct for the candidate's graph (within a half box tolerance). (1)

   If no graph drawn, 68 ± 1.

   Unit is not required; however zero marks are awarded for the correct value with incorrect unit.

   (c) Greater number/concentration/moles of hydrogen ions/H$^+$ (1)

   OR

   more H$^+$ ions.

2. (a) (i) Addition (1)

(ii) (1)

(b) Correctly drawn **full** structural formula for ethene. (1)

3. (a) As the percentage of carbon increases the heat content also increases. (1)

OR

As the percentage of carbon decreases the heat content also decreases.

OR

The heat content increases as the percentage of carbon increases.

OR

The heat content decreases as the percentage of carbon decreases.

(b) 46·67/46·7/47 (%) (3)

Partial marking:

$GFM = 120$ (1)

$$\frac{56}{candidate's\ GFM} \times 100$$ (1)

This step on its own is worth 2 marks if the candidate's gfm is 120.

Calculation of final answer using the relationship

$\%\ by\ mass = \frac{m}{GFM} \times 100$ (1)

(Unit is not required, however a maximum of 2 marks can be awarded for the correct value with incorrect unit.)

4. (a) Hydrogen and carbon. (1)

(b) Methylpropane (1)

OR

2-methylpropane

(c) Stronger and intermolecular forces. (1)

(d) 150 to 154 °C inclusive. (1)

(Unit is not required; however zero marks are awarded for the correct value with incorrect unit.)

5. (a) Sodium azide, potassium nitrate and silicon dioxide. (1)

(b) Potassium oxide. (1)

(c) SiO_2 (1)

(d) 44 (litres) (1)

(Unit is not required; however zero marks can be awarded for the correct value with incorrect unit.)

6. (a) (i) Two (1)

(ii) 10·8 (2)

11 with working (2)

Partial marking:

$$\frac{(10 \times 20) + (11 \times 80)}{100}$$ (1)

OR

$$\frac{10 \times 20\% + 11 \times 80\%}{100}$$ (1)

(Unit is not required; however a maximum of 1 mark can be awarded for the correct value with incorrect unit.)

(b) $^{14}_{6}C$ (1)

7. (a) (Ionic) Lattice (1)

(b) (i) Ions are free to move. (1)

(ii) Oxidation (1)

(iii) Allows the product(s) to be identified. (1)

OR

To make sure that only one product is produced at each electrode.

OR

To separate the strontium from the chlorine.

8. This is an open-ended question.

1 mark: The student has demonstrated a limited understanding of the chemistry involved. The candidate has made some statement(s) which is/are relevant to the situation, showing that at least a little of the chemistry within the problem is understood.

2 marks: The student has demonstrated a reasonable understanding of the chemistry involved. The student makes some statement(s) which is/are relevant to the situation, showing that the problem is understood.

3 marks: The maximum available mark would be awarded to a student who has demonstrated a good understanding of the chemistry involved. The student shows a good comprehension of the chemistry of the situation and has provided a logically correct answer to the question posed. This type of response might include a statement of the principles involved, a relationship or an equation, and the application of these to respond to the problem. This does not mean the answer has to be what might be termed an "excellent" answer or a "complete" one.

9. (a) (i) (A molecule that contains) carbon to carbon double bond. (1)

OR

C=C

(ii) Bromine (solution)(water)/Br_2 decolourised/discolourised/goes colourless. (1)

(b) 472·8/473 (kJ) (3)

Partial marking:

Using $cm\Delta T$ with $c = 1·97$ (1)

To be awarded this concept mark, candidates do not specifically need to write $cm\Delta T$. The concept mark is awarded for using this relationship with three values, one of which must be 1·97.

For values

1·5 (kg) **and** 160 (°C) (1)

A further mark can be awarded for arithmetical follow through to the candidate's answer **only if the mark for the $cm\Delta T$ concept has been awarded.** (1)

3 marks can be awarded for 472800 or 473000 J. **However the unit (Joules/J) must be given.**

(Unit is not required if answer given in kilojoules, however a maximum of 2 marks can be awarded for the correct value with incorrect unit.)

10. (a) (i) N_2 + $3H_2$ ⇌ $2NH_3$ **(1)**

    (ii) The reaction (it) is reversible **(1)**

       OR

       the reaction (it) occurs in both directions

       OR

       the reaction (it) reaches (or is at) equilibrium.

  (b) Diagram showing nitrogen with three hydrogen atoms:

    each of the three overlap areas must have two electrons in or on overlap area. **(1)**

    Either the nitrogen or all three hydrogen symbols must be shown.

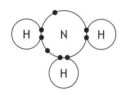

  (c) Ostwald **(1)**

  (d) Ammonium nitrate **(1)**

11. (a) From Mg to Cu through meter/wire. **(1)**

  (b) Completes the circuit/cell **(1)**

    OR

    allows ions to flow/move/transfer (between the two beakers)

    OR

    provide ions to complete the circuit/cell.

  (c) $Cu^{2+}(aq) + 2e^- \longrightarrow Cu(s)$ **(1)**

  (d) Insoluble **(1)**

    OR

    Solubility less than 1 g l^{-1}

12. (a) (Alpha particles) they **(1)**

    • cannot penetrate the paper

    • cannot pass through paper

    • are stopped by the paper

    • are absorbed by the paper

    • can **only** pass through air.

  (b) 14·8 (years) **(2)**

    Partial marking:

    1 mark can be awarded for **either:**

    • 4 half-lives

    OR

    • number of years correctly calculated for an incorrect number of half-lives (provided the working supports the number of half-lives).

    (Unit is not required; however a maximum of 1 mark can be awarded for the correct value with incorrect unit.)

  (c) Increases **(1)**

    Stays the same

13. (a) (i) Carboxyl **(1)**

    (ii) 134(g) **(1)**

       (Unit is not required; however 0 marks can be awarded for the correct value with incorrect unit.)

  (b) Any correct statement linking acidity to the position of the halogen. **(1)**

    e.g.

    The acidity (of the carboxylic acids) decreases going down the group.

    OR

    As you go (up) from iodine to fluorine the acidity increases.

    OR

    The one at the top (of the group) has the highest acidity.

    OR

    The one that has the lowest acidity is at the bottom (of the group).

  (c) A correct shortened or full structural formula for **(1)**

    4-methylpentanoic acid

    e.g.

    $CH_3CH(CH_3)CH_2CH_2COOH$

    $HOOCCH_2CH_2CH(CH_3)CH_3$

    OR

    mirror images.

    Accept CH_3 for branch in a full structural formula.

14. (a) C **(1)**

    OR

    0·24

  (b) (i) Pipette **(1)**

    (ii) (2 and 4) They are concordant. **(1)**

       OR

       They are within 0·2 (cm^3).

       OR

       They are within 0·1/0·1 apart.

       OR

Titration 1 and 3 or the other two are not concordant or not within 0·2 of each other.

(iii) 0·000161 (mol/mole/moles) (1)

OR

$1·61 \times 10^{-4}$

(Unit is not required; however 0 marks can be awarded for the correct value with incorrect unit.)

15. (a) Diamond(s) anvil cell (1)

(b) (i) 1·9 million (atmospheres) (1)

OR

1 900 000 (atmospheres)

OR

$1·9 \times 10^{6}$

(Unit is not required; however 0 marks can be awarded for the correct value with incorrect unit.)

(ii) The bulb would not light/would turn off. (1)

(c) $Fe_2O_3 \rightarrow O_2 + Fe_5O_7$ (1)

16. (a) (i) Homologous (series) (1)

(ii) $C_nH_{2n+1}SH/C_nH_{2n+1}HS$ (1)

OR

C_nSH_{2n+2}

OR

$C_nH_{2n+2}S$

OR

$C_nH_{2(n+1)}S$

(b) Sulfur (mon)oxide (1)

Sulfur dioxide

Sulfur trioxide

(c) 960 (g) (3)

Partial marking:

Both *GFMs* 32 and 48 (1)

OR

correct number of moles of methanol (20 moles). (1)

Correct application of the relationship between moles and mass. (1)

This could be shown by:

Method A

moles CH₃SH × candidate's GFM CH₃SH

OR

Method B

by working showing correct proportionality.

$640 \leftrightarrow \dfrac{candidate\ GFM\ CH_3SH}{candidate\ GFM\ CH_3OH} \times 640$

Where the candidate has been awarded the mark for the correct application of the relationship between moles and mass, a further mark can be awarded for correct follow through to a final answer. (1)

OR

Any other valid method accepted.

Unit is not required, however a maximum of 2 marks can be awarded for the correct value with incorrect unit.

Award zero marks if candidate's working does not use methanol.

17. **1 mark:** The student has demonstrated a limited understanding of the chemistry involved. The candidate has made some statement(s) which is/are relevant to the situation, showing that at least a little of the chemistry within the problem is understood.

2 marks: The student has demonstrated a reasonable understanding of the chemistry involved. The student makes some statement(s) which is/are relevant to the situation, showing that the problem is understood.

3 marks: The maximum available mark would be awarded to a student who has demonstrated a good understanding of the chemistry involved. The student shows a good comprehension of the chemistry of the situation and has provided a logically correct answer to the question posed. This type of response might include a statement of the principles involved, a relationship or an equation, and the application of these to respond to the problem. This does not mean the answer has to be what might be termed an "excellent".

NATIONAL 5 CHEMISTRY 2019

Section 1

Question	Answer	Mark
1.	C	1
2.	B	1
3.	C	1
4.	D	1
5.	B	1
6.	D	1
7.	A	1
8.	B	1
9.	A	1
10.	B	1
11.	D	1
12.	A	1
13.	D	1
14.	C	1
15.	A	1
16.	B	1
17.	B	1
18.	D	1
19.	C	1
20.	C	1
21.	C	1
22.	A	1
23.	D	1
24.	C	1
25.	B	1

Section 2

1. (a) (Covalent) Network (1)

    (b) (i) Isotope(s) (1)

        (ii) Different number of neutrons

           OR

           One atom has 5 neutrons and one atom has 6 neutrons (both numbers must be given)

           OR

           More/less neutrons (1)

        (iii) $11/^{11}B/^{11}_{5}B$ (1)

    (c) 120 (g) (1)

2. (a) Hydroxyl (1)

    (b) Enzyme (1)

    (c) Calcium oxalate (1)

    (d) 90 (g) (1)

3. (a) Must show all outer electrons (1)

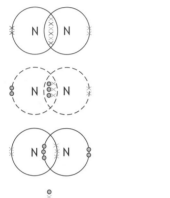

    Accept cross or dot or e or e- to represent electrons or a mixture of these.

    Accept petal diagram.

    If inner electrons are shown they must be correct.

    (b) Blue/purple (1)

       Any colour that clearly represents an alkaline pH is acceptable. (1)

    (c) (i) As the temperature increases yield/% decreases. (1)

       As the temperature decreases the yield/% increases.

       The yield/% increases as the temperature decreases.

       The yield/% decreases as the temperature increases.

       (ii) For appropriate format: points (not bars) (1)

       The axis/axes of the graph has/have suitable scale(s). For the graph paper provided within the question paper, the selection of suitable scales will result in a graph that occupies at least half of the width and half of the height of the graph paper. (1)

       The axes of the graph have suitable labels and units. (1)

       All data points plotted accurately with a line of best fit (smooth curve/straight line) drawn. (1)

    (d) (i) Electrolysis and lithium nitride (1)

       Both required for 1 mark

       (ii) Arrow from water at top to water at bottom OR to the vertical line beneath the 'lithium nitride' box. (1)

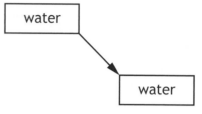

       Direction of arrow must be shown.

       OR

       Arrow from lithium hydroxide at bottom to lithium hydroxide OR heat at top OR to the vertical line between lithium hydroxide and heat.

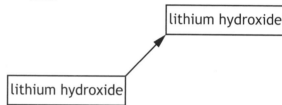

4. (a) Nucleus (1)

    (b) (i) The time for half of the mass/(radio)activity/nuclei to decay. (1)

       Radiation acceptable in place of radioactivity.

       (ii) $87.5\%/0.875/\frac{7}{8}$ (3)

       Partial marking

       3 half-lives (1)

       OR

       Correct percentage or fraction left for a correct number of half-lives. (2)

       OR

       Correct percentage or fraction left for an incorrect number of half-lives. Working must be shown. (1)

       OR

       Correct percentage or fraction decayed with incorrect number of half-lives. Working must be shown. (2)

       (iii) When an iodine-131 solution is diluted, (1)

       the half-life gets longer / stays the same / gets shorter

5. (a) Bromine/Br_2 decolourised (1)

       OR

       bromine/Br_2 goes colourless

       Accept bromine/bromine water/bromine solution but do not accept bromide or Br.

    (b) (i) Hydrogenation (1)

       (ii) Chlorine/Cl_2 (1)

       (iii) Polypropene/Poly(propene) (1)

       Accept polypropylene

5. (c) (i) 140 (g)   (3)

        Partial marks

        Both GFMs

        ie 28 and 82   (1)

        OR

        Moles of cyclohexene

        ie (410 ÷ 82) = 5 mol   (1)

        Award 0 marks if the candidate's working does not use cyclohexene.

        1 concept mark for either:

        $410 \times \dfrac{\text{Candidate's GFM of ethene}}{\text{Candidate's GFM of cyclohexene}}$ (1)

        OR

        Moles of cyclohexene × Candidate's GFM of ethene   (1)

        Where the candidate has been awarded either concept mark, a further mark can be awarded for correct follow through to a final answer   (1)

    (ii) (Cyclopentene) has smaller/weaker/less forces (of attraction)   (1)

        Between molecules or mention of intermolecular attractions   (1)

        If neither of these two points are given, a maximum of 1 mark can be awarded for – cyclopentene is smaller/has fewer carbons or hydrogens or atoms/smaller carbon chain.

6. This is an open-ended question.

    1 mark: The student has demonstrated a limited understanding of the chemistry involved. The candidate has made some statement(s) that is/are relevant to the situation, showing that at least a little of the chemistry within the context is understood.

    2 marks: The student has demonstrated a reasonable understanding of the chemistry involved. The student makes some statement(s) that is/are relevant to the situation, showing that the context is understood.

    3 marks: The maximum available mark would be awarded to a student who has demonstrated a good understanding of the chemistry involved. The student shows a good comprehension of the chemistry of the situation and has provided a logically correct answer to the question posed. This type of response might include a statement of the principles involved, a relationship or an equation, and the application of these to respond to the context. This does not mean the answer has to be what might be termed an "excellent" answer or a "complete" one.

7. (a) They have the same general formula

        AND

        similar/same chemical properties

        Both required for 1 mark

    (b) (i) Alkane   (1)

        (ii) $C_{35}H_{72}$

            Numbers must be smaller than symbol or subscript

            Symbols must be correct

    (c) Accept 391 to 394 inclusive.   (1)

8. (a) Supernova (explosions)   (1)

    (b) Beryllium, aluminium, silicon and oxygen Be/Al/Si/O

        All 4 required for 1 mark

    (c) $BeCl_2 + K \longrightarrow Be + KCl$ (1)

        Equation does not need to be balanced but if balanced it must be correct:

        $BeCl_2 + 2K \rightarrow Be + 2KCl$

    (d) Reduction   (1)

    (e) $_{0}^{1}n$ (1)

9. (a) Exothermic/exothermal   (1)

    (b) (i) 0·05 (kg)   (3)

        Partial marking:

        Using the correct concept of

        $m = \dfrac{E_h}{c\Delta T}$ with

        c = 4·18   (1)

        (To be awarded this concept mark candidates do not have to specifically write $\dfrac{E_h}{c\Delta T}$.)

        For values

        8·36 and 40   (1)

        A further mark can be awarded for arithmetical follow through to the candidate's answer **only if the mark for the concept has been awarded.**   (1)

        8360 and 4180 can be used together in the calculation.

        8360 and 40 can be awarded 1 mark provided c = 4180.

        (ii) Copper can: copper is a better conductor (of heat) (than glass).   (1)

            Heat shield: prevent/less heat loss (to the surroundings) or insulate the experiment.   (1)

            These two marks are independent of each other.

            A comparison of energy transfer to the water must be made however 0 marks are awarded for a comparison in terms of easier/quicker/faster.

            Link must be made to heat/energy

    (c) (i) If the OH/functional group is at the end/start/carbon number 1 (of the alcohol), an aldehyde is produced.

        OR

        If the OH/functional group is not at the end/in the middle (of an alcohol), a ketone is produced.   (1)

        (ii)   (1)

10. (a) Ion bridge/salt bridge   (1)

    (b) (i) From right to left either on the wire or nearer the wire/voltmeter than the ion-bridge.   (1)

        (ii) Oxidation   (1)

        (iii) $2Fe^{3+} + 2I^- \rightarrow 2Fe^{2+} + I_2$ (1)

            Accept correct multiples.

(c) Conducts (electricity) (1)

Additional correct properties do not negate.

Accept a correct description of electrical conductivity e.g. electrons free to move or has delocalised electrons.

11. (a) $Ca^{2+}CO_3^{2-}$ (1)

(b) Propanoic acid. (1)

Accept propaneoic or propionic acid.

(c) until…

no more solid reacts/until it no longer reacts

solid remains/is left (at the bottom of the beaker)

a gas is no longer produced no more fizzing/bubbling

no more calcium carbonate reacts/calcium carbonate left (at the bottom)

neutral/neutralised **with a description of testing pH.** (1)

(d) 1st technique — Filtration

2nd technique — Evaporation

Both required in correct order for 1 mark. (1)

Accept a correct description of filtration.

Accept heat it/warm it/boil it/leave by a window/leave for some time/distillation.

12. (a) (i) Standard (solution) (1)

(ii) 21·2 or 21 (g) (3)

Partial marks can be awarded for a maximum of two of the following three steps:

1 mark for correctly calculating the number of moles of sodium carbonate

i.e. $n = CV = 1 \times 0·2 = 0·2$ mol (1)

1 mark for correctly calculating the GFM of sodium carbonate

i.e. GFM = 106 (1)

1 mark for calculating the mass of sodium carbonate

i.e. m = n × GFM using candidate's calculated moles of sodium carbonate and candidate's calculated GFM (1)

(b) (i) Any from:

Move burette/sodium carbonate (reading) to eye level.

Reduce distance between flask and burette.

Use white tile/paper.

Clamp the burette towards the centre. (1)

(ii) To show the end-point/point of neutralisation/completion of the reaction. (1)

(iii) Concordant (1)

(iv) 1·5 (mol l^{-1}) (3)

Partial marks can be awarded for a maximum of two of the following three steps:

Method A

1 mark for the correct number of moles of sodium carbonate.

ie $n = CV = 1 \times 0·015 = 0·015$ mol (1)

1 mark for calculating the moles of hydrochloric acid by correctly applying the molar ratio ie 2 × candidate's calculated number of moles of Na_2CO_3 (1)

0·03 mol HCl on its own (2)

1 mark for calculating the concentration of the hydrochloric acid. ie C = n/V using candidate's calculated number of moles of HCl and 0·02 (1)

Method B

$$\frac{C_1 \times 20}{2} = \frac{1 \times 15}{1}$$ (1)

$C_1 \times 20 = 30$ or $C_1 \times 10 = 15$ (1)

(this step on its own gets 2 marks)

OR

$$\frac{C_1 \times 0.02}{2} = \frac{1 \times 0.015}{1}$$ (1)

$C_1 \times 0·02 = 0·03$ (1)

(this step on its own gets 2 marks)

Award 0 marks if the candidate's working makes it clear that they have not used sodium carbonate.

Accept mol l^{-1} or mol/l but not mol/l^{-1} or mol^{-1} or mol l

If candidate does not attempt to calculate the number of moles of Na_2CO_3 then partial marks can be awarded for applying the molar ratio using 1 mol of sodium carbonate requires 2 moles of hydrochloric acid and/or using 2 moles of hydrochloric acid and V = 0·02.

ie 1:2 (1)

2/0·02 = 100 (1)

For method B using relationship shown in the data book.

13. **1 mark:** The student has demonstrated a limited understanding of the chemistry involved. The candidate has made some statement(s) that is/are relevant to the situation, showing that at least a little of the chemistry within the context is understood.

2 marks: The student has demonstrated a reasonable understanding of the chemistry involved. The student makes some statement(s) that is/are relevant to the situation, showing that the context is understood.

3 marks: The maximum available mark would be awarded to a student who has demonstrated a good understanding of the chemistry involved. The student shows a good comprehension of the chemistry of the situation and has provided a logically correct answer to the question posed. This type of response might include a statement of the principles involved, a relationship or an equation, and the application of these to respond to the context. This does not mean the answer has to be what might be termed an "excellent" answer or a "complete" one.